Automotive Milestones

Englisch für Fahrzeugberufe

von
Dr. Wolfgang Schäfer (Hrsg.)
Florian Bär
Annely Humphreys
Jason Humphreys
Linda Keller
Mary Schäfer

Ernst Klett Verlag
Stuttgart · Leipzig

Automotive Milestones

Englisch für Fahrzeugberufe

Autoren: Dr. Wolfgang Schäfer (Hrsg.), Florian Bär, Annely Humphreys, Jason Humphreys, Linda Keller, Mary Schäfer

Werkübersicht:

Schülerbuch, 978-3-12-808281-3
Milestones Workbook für alle technischen Fachbände mit Prüfungsvorbereitung KMK-Fremdsprachenzertifikat und CD-ROM, Übungsbegleitheft für alle technischen Berufe, 978-3-12-808285-1
Milestones Lehrerhandbuch für technische Berufe inkl. Digitalem Lehrer-Service mit Medien-DVD-ROM + Lehrer-Audio-CDs (6), 978-3-12-808287-5

1. Auflage 1 9 8 7 6 5 | 26 25 24 23 22

Alle Drucke dieser Auflage sind unverändert und können im Unterricht nebeneinander verwendet werden.
Die letzte Zahl bezeichnet das Jahr des Druckes.

Im Lehrwerk befinden sich ausschließlich fiktive Internet-Adressen, die deshalb auch mit ww#. beginnen anstatt wie üblich mit www.
Die im Buch abgedruckten Mediencodes führen zu interaktiven Zusatzübungen und Hörverständnistexten auf www.klett.de. Die Mediencodes leiten ausschließlich zu optionalen Unterrichtsmaterialien, sie unterliegen nicht dem staatlichen Zulassungsverfahren.

Projektleitung: Matthias Rupp, Karin Altrichter
Redaktion: Dr. Birgit Reinel, Tübingen; Volker Wendland, Helen Smyth
Herstellung: Angelika Lindner, Sarah Ganser

Satz und Gestaltung: Satzkiste, Stuttgart
Umschlaggestaltung: Projektteam des Verlages
Reproduktion: Meyle + Müller Medien-Management, Pforzheim
Druck: Himmer GmbH Druckerei, Augsburg

Printed in Germany
ISBN 978-3-12-808281-3

9 783128 082813

Automotive Milestones bietet Ihnen praxisorientiertes Fachenglisch für Fahrzeugberufe an Berufsschulen, Berufsfachschulen, Fachschulen sowie für technisch ausgerichtete Englischkurse in der Erwachsenenbildung.

Automotive Milestones im Überblick:

- Konsequente Berücksichtigung neuester Lehrpläne (Lernfelder) sowie durchgängiges Sprachkompetenztraining nach dem Gemeinsamen Europäischen Referenzrahmen (Sprachstufen A2 – B2).
- 21 kompakte 4-seitige Module: Fachmodule mit Schwerpunkt auf Fachenglisch wechseln sich mit Modulen (A – H) ab, in denen der Schwerpunkt auf der beruflichen Kommunikation liegt.
- Schnelle Orientierung im Buch durch Übersicht der Lernziele jeweils am Modulanfang.
- Die Kompetenzen auf einen Blick: Viele Aufgaben bereiten bereits gezielt auf die Anforderungen des KMK-Zertifikats hin. Diese Aufgaben sind mit den Kürzeln R (Rezeption), P (Produktion), I (Interaktion) und M (Mediation) gekennzeichnet.
- Typische Sprachhandlungssituationen im Berufsalltag werden multimedial mit Audio- und Videounterstützung (u.a. Originalvideos der BBC) trainiert.
- Zahlreiche Partner-, Gruppen- und Internetaufgaben werden durch Texte und Aufgaben zur Arbeitssicherheit *(Safety first)* abgerundet.
- In den Modulen befinden sich hilfreiche Verweise auf die nach beruflichen Situationen geordneten Redewendungen *(Phrases)*.
- Unter der Überschrift *Understanding technical English* befinden sich Hinweise und Aufgaben zum richtigen Umgang mit englischem Fachvokabular.
- Möglichkeit zur Binnendifferenzierung durch anspruchsvollere Aufgaben sowie Aufgaben zur Vorbereitung auf das KMK-Fremdsprachenzertifikat auf verschiedenen Stufen.
- Auf den Fachwortschatz in den jeweiligen Modulen kann über die *Word bank* (auf jeder Doppelseite) direkt zugegriffen werden.
- Online-Grammatiktraining über die Milestones-Codes 🌐 3u5mw9 in den Modulen A – H.
- Vertieftes Grammatik- und Hörverstehenstraining im Workbook mit Prüfungsvorbereitung auf das KMK-Fremdsprachenzertifikat, inklusive CD-ROM mit allen technischen Milestones-Fachbänden- und Workbook-Audios (MP3).

Weiterführendes Material

Milestones Workbook für alle technischen Fachbände mit Prüfungsvorbereitung KMK-Fremdsprachenzertifikat und CD-ROM

Lehrerhandbuch für alle technischen Milestones-Fachbände mit 6 Audio-CDs + Digitaler Lehrer-Service (Lehrer-Software mit multimedialen Schülerbüchern, Audios, Videos, Lösungen, Übungsmaterial, Übungsgenerator, Dokumentenpool mit veränderbaren Materialien)

Videos, Anhang, Symbole ▶

Üben interaktiv: Online Grammatiktraining

Modul A: 3u5mw9	**Modul E:** 5zw8jx
Modul B: gc2x9w	**Modul F:** pf5r3n
Modul C: x3gj95	**Modul G:** 3b9c4h
Modul D: cy8yn5	**Modul H:** 842j42

Die **Module A – H** legen den Schwerpunkt auf die Skills zu Business Communication, die im Workbook intensiv trainiert werden.

Video lounge (Die Videos und die Videoskripte befinden sich im Lehrerhandbuch.) 92

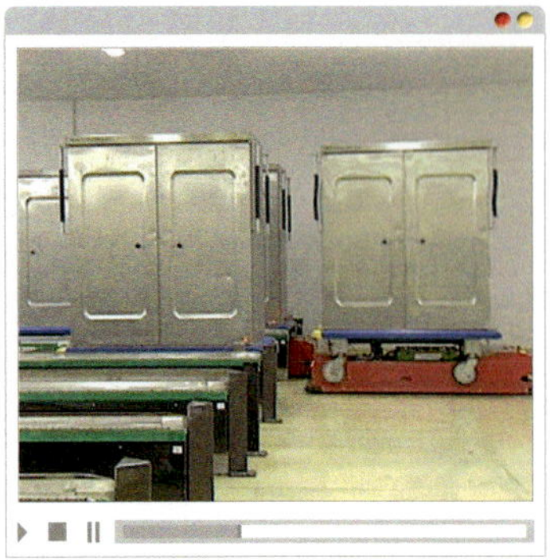

Appendix

Umschlagseiten

Modulelemente

Nach diesem Modul	Lernziele
WORD BANK	Fachvokabular
UNDERSTANDING TECHNICAL ENGLISH	Hinweise und Aufgaben zu englischem Fachvokabular
SAFETY FIRST ⚠	Aufgaben zur Arbeitssicherheit
→ **PHRASES:** Telephoning	Verweis auf *Phrases (Telephoning)* im Anhang

Symbole

◎ **A 1.2**	Audioverweis mit Tracknummer (CD1 – Track 2)
P, M, I, R	Produktion, Mediation, Interaktion, Rezeption
(Globus-Symbol)	Aufgabe mit Internetrecherche
(Dreieck-Symbol)	Anspruchsvollere Aufgabe *(advanced)*
(Personen-Symbole)	Partner- und Gruppenaufgaben
(Globus-Symbol) 3u5mw9	Grammatiktraining online über www.klett.de
KMK II	Aufgabe zur Vorbereitung auf die Prüfung des KMK-Fremdsprachenzertifikats (Niveaustufe II)

01

NACH DIESEM MODUL:

- ✓ kenne ich grundlegende englische Begriffe für Fahrzeugklassen und Typen,
- ✓ kann ich Teile eines Fahrzeugs auf Englisch benennen und Fahrzeuge auf Englisch beschreiben.

1

2

3

4

Introduction to vehicles

Nearly everybody in Europe owns or uses a car – cars of different manufacturers, different sizes and powered by different types of engines. For some people their car is a status symbol; for others it is simply a way of getting from one place to the next. And then there are the environmentalists, who argue that cars are bad for the environment and a danger to our future. No matter what you think about cars, in this business you will need the English words and expressions to talk about them with confidence.

1 **Look at the photos and answer the following questions.**

1. Would you like to own one of the cars pictured above? Why or why not?
2. Which car do you like best? Give reasons.
3. Can you think of a possible classification for each of the cars?
4. What brands do you prefer? Explain why you like them.
5. What models does your favourite car manufacturer produce?

2 **Create a word web about cars and compare your results in class.**

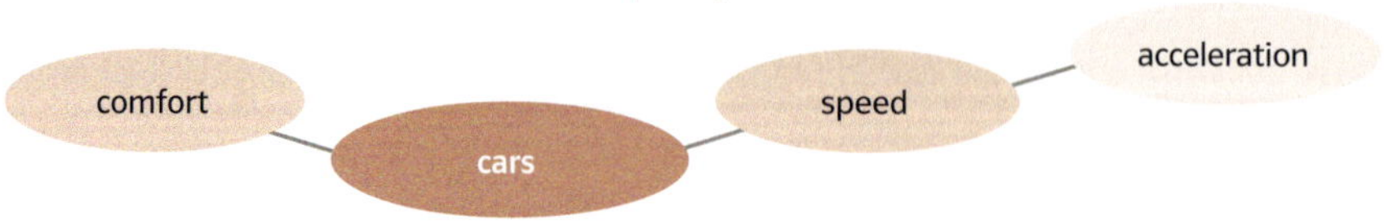

A | Classifying cars

A common way of classifying cars is according to size, performance and price. What car is best for you depends not only on your budget, but also on the intended purpose.

1 Match the types of cars to their definitions. Then match the photos on page 8 to the types of car in the list below.

1. microcars 2. superminis 3. compact executives 4. large family cars 5. compact MPV cars 6. small family cars 7. executive cars 8. luxury cars 9. sports cars 10. SUVs	a. quite inexpensive, low maintenance costs b. space for a small family c. space for a bigger family and their luggage; comfortable to drive d. spacious cars with powerful engines, cruise control, in-car computer and automatic start-stop function e. saloon cars with powerful V6 and V8 engines, the latest technology and numerous luxurious features f. usually two seats, two-door automobile designed for high speed driving, good handling, minimum weight and high performance g. popular with families needing extra room and more versatility than a supermini can offer h. group of cars that are the smallest on the road with a maximum of two seats i. usually equipped with four-wheel drive for on- or off-road driving j. often bought by businesses for people who have to represent their company

P **2** Find examples of the types of cars mentioned above. Can you think of other car classes that are not listed? Choose two types of cars and compare them.

R **3** Listen to the radio broadcast and answer the questions.
A 4.1

1. What classes of cars are mentioned? Give examples of each class.
2. What does the abbreviation MPV stand for?
3. What is a crossover?
4. What example is mentioned for a coupé car?
5. Why do people buy four-wheel drive cars?

R **4** Listen again and find suitable cars for the following target groups. Give reasons for your choice.
A 4.1

single, high income • family, four children • businessman, likes climbing

5 Check the Internet for car manufacturers in the UK. Choose one and present some models to your class.

WORD BANK

brand – Marke
cruise control – Tempomat, Geschwindigkeitsregelanlage
engine – Motor
feature – *(hier:)* Ausstattung
four-wheel drive – Allradantrieb
in-car computer – Bordcomputer
maintenance costs – Wartungskosten
manufacturer – Hersteller
performance – Leistung
vehicle – Fahrzeug

WORD BANK

boot (BE) / trunk (AE) – Kofferraum
running costs – Unterhaltskosten

B | Parts of a vehicle

1 **UNDERSTANDING TECHNICAL ENGLISH**

Match the English terms with the numbers in the photo.

a. bonnet (BE)/hood (AE)	g. radiator grille (BE)/grill (AE)
b. bumper	h. wing (BE)/fender (AE)
c. boot (BE)/trunk (AE)	i. roof
d. headlight	j. door handle
e. tyre (BE)/tire (AE)	k. wipers
f. windscreen (BE)/windshield (AE)	l. wing mirror (BE)/side mirror (AE)

2 **Describe your own or your favourite car using the words in exercise 1.**

3 **Fill in the missing words using the illustration below.**

a. The ? is where you sit while driving a car.
b. The ? is something that you press with your foot to make the car go faster.
c. The ? is an instrument in a vehicle that shows how fast you are driving.
d. The ? makes a vehicle slow down or stop.
e. The ? helps you to get to your destination.
f. The ? is a metal rod that you move to control the gears of a vehicle.
g. The ? is the part in front of the driver where the controls are located.
h. You have to press the ? with your foot in order to change gear.
i. The small compartment in front of the passenger seat is called ?.

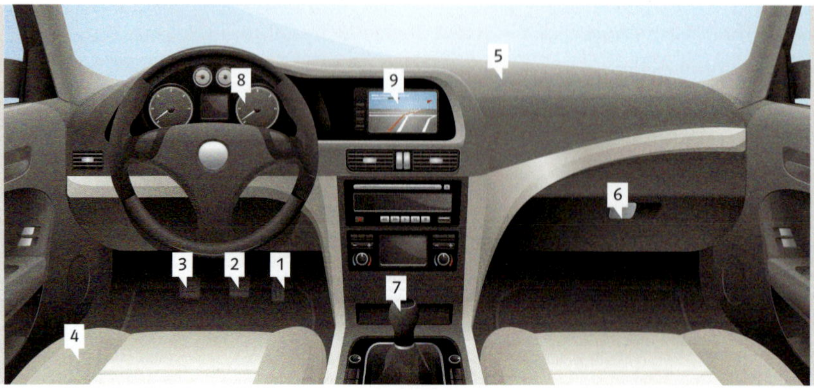

1. accelerator (BE)/ gas pedal (AE)
2. brake
3. clutch
4. driver's seat
5. dashboard
6. glove compartment
7. gear stick
8. speedometer
9. GPS

C | Choosing the right car

M **1** Read the following text and sum it up in German.

ww#.car_sales.com

What car? **Advice on making the right choice**

There is a wide range of cars on the market, and choosing the right one for you can be difficult. The best thing to do is to decide what you need from your car and how much you can afford to spend. Consider the following issues:

Your budget
Here you need to know not only the price of a new car, but also what it costs to run the car (e.g. fuel and maintenance). Speak to your dealer for information on servicing costs.

Fuel consumption
Basically, the less fuel your car uses, the better. Some modern economical cars have an average fuel consumption of around 7 liters per 100 km. The majority of these have diesel engines, but small petrol engines can run them close.
Some cars, such as hybrids are getting more economical all the time.

Taxes
Since 2009 road tax has been based not only on the cubic capacity of the engine, but also on how much carbon dioxide (CO_2) a car emits. So the greener your car is, the cheaper the tax.

What's the car for?
Is the car for driving to work every day, going on holiday or just for fun? If you're looking to buy a car for everyday life then you want something relaxing and comfortable to drive, with plenty of extras such as air conditioning, heated seats or a good hi-fi system. And if you're not going to be driving through rough terrain, you probably don't need four-wheel drive or sports suspension.

How many passengers?
Additional space can be a waste and bigger cars often cost more to run than smaller ones, so think about how many people will you usually have in the car.

Petrol or diesel?
If you buy a car for short trips it's probably a good idea to buy a model with a small petrol engine – they're generally cheaper than cars with a diesel engine. However, for long trips and driving on the motorway, a diesel makes sense.

The boot
A huge boot means that the car can take you and your passengers on holiday or accommodate pets. It's worth making sure how much space a car has with its rear seats in place, and folded down for more luggage room.
A small city car usually has just enough space in its boot for shopping bags but it is usually not suitable to go on holiday with four of you.

Specific tasks?
If you need to carry out specific tasks, make sure the car is up to the job. For example, if you tow a trailer, check the official maximum trailer weight, which must never be exceeded.

P **2**
KMK I

Sie wollen ein neues Auto kaufen und finden im Internet ein englischsprachiges Forum, in dem sich die Mitglieder Hilfestellung für den anstehenden Fahrzeugkauf geben. Verfassen Sie einen Beitrag und gehen Sie dabei auf folgende Punkte ein.

Reparatur des alten Fahrzeugs zu teuer • Neuwagen oder Gebrauchtwagen? • sparsam, aber mindestens 90 kw • auf alle Fälle Schaltgetriebe, kein Automatikgetriebe • Platz für 4 Personen • viel Gepäckraum • max. 18.000 Euro • europäische Marke

WORD BANK

air conditioning – Klimaanlage
automatic transmission – Automatikgetriebe
bonnet (BE)/hood (AE) – Motorhaube
bumper – Stoßstange
cubic capacity – Hubraum
door handle – Türgriff
fuel consumption – Kraftstoffverbrauch
gear – Gang
headlight – Scheinwerfer
manual transmission – Schaltgetriebe
radiator grille (BE)/grill (AE) – Kühlergrill
rear seat – Rücksitz
suspension – Federung, Aufhängung
tyre (BE)/tire (AE) – Reifen
windscreen (BE)/windshield (AE) – Windschutzscheibe
wing mirror (BE)/side mirror (AE) – Außenspiegel
wiper – Scheibenwischer

A

NACH DIESEM MODUL:

- ✓ weiß ich, wie man sich auf Englisch begrüßt,
- ✓ weiß ich, wie ich mich selbst und andere einander auf Englisch vorstelle.

1

2

3

4

Meeting people

In the world of work you will often meet new people and it will be necessary to introduce yourself. In all industries developing relationships with colleagues and customers is the key to success. Not only is it important that you are polite, confident and friendly, but also that you know what to say, what information to include, and what the differences between formal and informal situations are.

1 **Match the parts of the dialogues to form short conversations. Start with Speaker A.**

Speaker A	Speaker B
1. Morning, Tom. That was a great game last night, wasn't it? 2. Hi guys, my name's Natalie. 3. Goodbye, Mr Bauer. See you next week. 4. It's nice to meet you, Simon.	a. Nice to meet you, too, Hannah. b. Goodbye, Mr Horvath. It was nice meeting you. c. Morning Chris. Yeah, it was amazing! d. Hi Natalie, I'm Mark. Welcome to the team.

2 **Now match the conversations above with the appropriate picture (1–4) and decide which are formal and which are informal.**

P **3** **Which other phrases do you know for greeting somebody and saying goodbye? Make a list.**

A | Saying "hello" and "goodbye" and giving your name

A trainee from Denmark has just arrived at Neureuther und Söhne in Kiel for a three-month internship.

R **1** ⓞ A 1.1 **Listen to the two dialogues and write down the names of the people and their positions in the company.**

R **2** ⓞ A 1.1 **Listen again and write down the phrases the people use for greeting and introducing one another.**

R **3** **Match the replies to the greetings and goodbyes in the two boxes.**

Formal greetings/goodbyes	Formal replies
1. Good morning. How are you today?	a. Good morning. I'm fine, thanks. And you?
2. Good night. Have a nice evening, Mr Dachs.	b. Goodbye, Mr Schäfer, and thanks for everything.
3. Goodbye, Mr Horvath. Have a nice journey home.	c. Good night, Ms Kistner, you too.
4. Good night, Silke.	d. Good night, Ms Hornbach.

Informal greetings/goodbyes	Informal replies
1. Hello, Mike. Haven't seen you for a long time.	a. Bye, Thomas. Give me a call, will you?
2. Lena, how are you this morning?	b. You, too, Florian. Thank you.
3. Morning. How was your weekend?	c. Fine, thanks. How about yours?
4. Bye-bye, Sevda.	d. I'm good, thanks. Yourself?
5. Have a nice weekend, Joe.	e. Yes, I was away on a trip last week.

P **4** **Make dialogues from the following prompts and present them to the class. Look at your lists of phrases from exercise 2 and check the info box on the next page.**

Example:
A: Hello, my name is Marina Kohler. I represent Toolcell in Leipzig.
B: Hello, Ms Kohler. I'm Amy Strong from Easytools in Bern. Nice to meet you.
A: Nice to meet you, too, Ms Strong.

Names: Denise Hofmann • Marina Kohler • Alice Campbell • Abel Kazich • Tanja Neuhäuser • Rudolfo Orvieto • Ursula Braun • Stefan Helm • Gerard Mathieu • your name

Job descriptions: I'm a trainee/intern • I work for … • I come from … • I represent … • I'm training to be a … • your job

Companies: Toolcell GmbH in Leipzig, Germany • Easytools in Bern, Switzerland • Tools Expert Pontypridd, Wales • DWT Electronics, Stuttgart • Smart EDV, Kiel • Xiang Piong Ltd, Hong Kong • Naturholz GmbH • your company

Situations: a first meeting with a new customer • a meeting with an existing customer/a colleague • a party • a company open day • a trade fair • a job fair for trainees • the first day in a new company

WORD BANK ⓞ

intern – Praktikant(in)

5 Introduce the people to each other according to the following structure. Use the prompts on the previous page.

→ **PHRASES:** Meeting people

A: Fragen Sie B und C, ob sie sich schon kennen. B und C antworten mit Nein.
A: Stellen Sie B und C einander vor.

B: Begrüßen Sie C freundlich.

C: Sie grüßen B zurück und stellen eine Frage zu Bs Firma

B: Sie geben eine Auskunft über Ihre Firma. Sie fragen C nach seiner Tätigkeit.

C: Sie beantworten Bs Frage.

INFO BOX: Meeting people

First meeting / Formal introduction

A: Hello, my name is Ines Sacher.
B: Nice to meet you, Ms Sacher. My name is Joshua Hamsung.
A: Nice to meet you, too, Mr Hamsung.
Women – married or not – are addressed as Ms, not Miss.

A: Mr Jacobs, this is Mr Sykes.
B: How do you do?
C: How do you do?
"How do you do?" is rarely used for first introductions.

A: I'm Frank Warren of Wood Tools in Manchester. Please call me Frank.
B: Pleased to meet you, Frank. My name is Joanne McNeill. Do call me Joanne.
In English-speaking countries first names are often used from the start of a business realtionship. In South America or Asia, however, you don't usually use first names right away.

I hope you had a good trip. /
I hope you had no trouble finding us. /
Welcome to our stand.
Phrases like these are used to break the ice after an introduction.

Second meeting

A: Good morning, Ms Sacher. Do you remember me? I'm Joshua Hamsung from Bell & Co in Chicago.
B: Good morning, Mr Hamsung. Nice to see you again.

Informal meetings

A: Hello Ines. Great to see you again. How are you?
B: Hi Joshua. Lovely to see you, too. I'm fine. And you?

B | Intercultural awareness

M **1** KMK II **Ihr Betrieb wird in Zukunft für einen amerikanischen Auftraggeber arbeiten. Ihre Vorgesetzte interessiert sich daher für folgenden Artikel aus einer englischen Fachzeitschrift. Sie bittet Sie, die wesentlichen Aussagen des Artikels stichpunktartig in übersichtlicher Form für sie auf Deutsch herauszuarbeiten.**

Intercultural awareness

Politeness, pleasantries and profitability

Americans are known to be direct, but they do like to take a few minutes to chat before getting down to work. They may ask after your health, or if your trip was pleasant, or about where you are from. This gives both parties a chance to form an impression of the other and establishes a basis on which to do business.

Americans, including the president, like to be called by their given name. This is not an invitation to be personal, as you would expect in Germany if someone suggested using the familiar "Du". It is a way to set a framework for working together. While it is usual in business, it would not be expected in dealing with a doctor, a teacher or a professor, and in any case, you should wait until it is offered: "Hello, I'm James Smith, vice-president in charge of sales. Call me Jim." "Pleased to meet you! I'm Walter König; please call me Walter." (Note that in the United States you state your *given* name and then your *family* name, and not the other way around, except perhaps on a form that you have to fill out.)

Americans don't like to waste time. They get to the point, in conversations as well as in conferences. Although Milton Berle, a comedian from the 1950s and 1960s, is quoted as saying: "A committee is a group of people that keeps minutes and wastes hours."

Americans are generally very polite, and they tend to smile much more than we are used to. An American tourist was quite upset that perfect strangers didn't smile back at her when she walked through the streets of a small town in Germany. On the other hand, a German woman in New York became quite tired of being asked, again by perfect strangers, the person at the cash register, the bus driver: "How are you doing?" and commanded to "Have a nice day!" These pleasantries are intended to show that you are noticed and respected. More than "Thanks!" is not expected.

P **2** **Check the Internet to see what advice people from other countries get to prepare for meetings in Germany. Make a list of ten useful tips.**

02

NACH DIESEM MODUL:

- ✓ kenne ich die englischen Begriffe für meine Werkzeuge,
- ✓ kann ich die Arbeiten mit ihnen auf Englisch ausdrücken,
- ✓ kann ich Standardsituationen in der Werkstatt auf Englisch bewältigen.

1

2

3

4

Tools in a car workshop

You will use a wide variety of tools in the workshop. When you repair cars you need common tools, such as screwdrivers, spanners, etc. and special purpose tools, such as a torque wrench or an adjustable wrench. When you are talking to English-speaking colleagues and customers, it is important to know what the tools are called in English.

1 Match the tools (a. – f.) to the correct categories (1. – 4.). Which of these tools are shown in the photos above?

	a. torque wrench
1. gripping tools	b. tyre pressure gauge
2. cutting tools	c. hacksaw
3. measuring tools	d. pliers
4. special automotive tools	e. multimeter
	f. lug wrench

2 Check the Internet and find other hand tools you often use in the workshop. List them under the four categories in exercise 1.

3 **UNDERSTANDING TECHNICAL ENGLISH**

Check out the words 'spanner' and 'wrench' in an English-German dictionary. Is there any difference in meaning? When are the terms 'spanner' and 'wrench' used?

A | Tools – Usage and safety

1 Match the tools with their definitions and usage.

1. ring spanner	a. synonym for 'hexagon socket key'
2. set of wrenches	b. spanner with a metal loop at the end
3. extension	c. device to make something longer
4. combination pliers	d. the handle we use for socket wrenches
5. chisel	e. used to loosen or tighten nuts and bolts
6. Allen key	f. tool used for shaping stone or metal
7. ratchet	g. pliers designed for many different purposes
8. punch	h. tool to mark a hole before you drill it

2 Find out what these tools are called in German.

file • grinder • grip pliers • side cutter • Phillips screwdriver • slot screwdriver • soft-faced hammer • tyre lever

I/M **3** Use the role cards on pages 104 and 105 to make a dialogue on the usage of tools.

4 SAFETY FIRST ⚠

The following safety tips are jumbled. Put them into the correct order.

1. use • you are not • Do not • tools • familiar with
2. and tangle-free • Keep • off the ground • power cords
3. power tools • off • or unplugging them • plugging • before • Turn
4. when • wet • Never • electrical tools • use • are • your hands
5. electrical • Use • tools • for • insulated • work
6. when • grinders • Wear • glasses • operating • safety
7. clean • area • Keep • and tidy • work • your
8. in the • a fire • and a first • Always • extinguisher • workshop • keep • aid kit

P **5** ⚠ Make a list of similar safety tips for your workshop. Present them in class.

R **6** You want to change the tyres on a car. Put the steps in the correct order.

a. Loosen the wheel nuts by using an impact driver or tyre lever.
b. Lower the jack slowly.
c. Label the removed wheel.
d. Before removing the wheels, secure the car against rolling away.
e. Jack up the car by using a lifting platform or a manual jack.
f. Tighten the wheel nuts to correct torque in a diagonal pattern.
g. Check the air pressure to ensure safe driveability and normal wear.
h. Remove the tyre.
i. Position the spare tyre over the wheel hub.

7 Tell your partner what your tool trolley looks like. Describe what is in it and make a rough sketch of each drawer. Use a dictionary if necessary.

WORD BANK

Allen key – Innensechskantschlüssel, Inbusschlüssel
bolt – (Schrauben-)Bolzen
chisel – Meißel
combination pliers – Kombizange
device – Gerät, Vorrichtung
extension – Verlängerung
hacksaw – Bügelsäge
hexagon socket key – Innensechskantschlüssel, Inbusschlüssel
impact driver – Schlagbohrer
jack – Wagenheber
jack up (v) – aufbocken, hochheben
lifting platform – Hebebühne
loop – Öse, Schlinge
loosen (v) – lösen, lockern
lug wrench – Radmutternschlüssel
nut – Schraubenmutter
pliers – Zange
punch – Körner
ratchet – Ratsche
ring spanner – Ringschlüssel
screwdriver – Schraubendreher
socket wrench – Steckschlüssel
spare tyre – Ersatzreifen
tighten (v) – festziehen, spannen
torque wrench – Drehmomentschlüssel
tyre lever – Reifenmontierhebel
tyre pressure gauge – Reifendruckmesser
wear – Abrieb, Verschleiß
workshop – Werkstatt

B | Types of tools

1 Look at the photos and say which tools they show: a pneumatic wrench, a lug wrench or a wheel balancing machine?

1

2

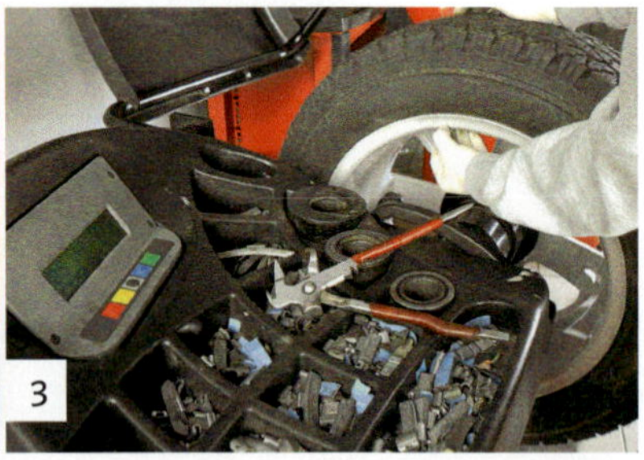
3

2 Copy the table and put in the words in the box.

hammer • power drill • chisel • column drill • wire strippers • centre lathe • angle grinder • impact driver • grinding machine • milling machine • tyre changing machine • spanner • torque wrench

Hand tools	Power tools	
	Manual power tools	*Machine tools*
lug wrench	pneumatic wrench	wheel balancing machine

P **3** Use the Internet or a dictionary and find more hand tools and power tools. Add them to your table and report back to the class.

P **4** Use the prompts below to make a dialogue.

A: What's the best tool for …?
B: Use/Take a(n) … for loosening/tightening/shortening/…
A: What tool(s) do I need for …?
B: I would recommend taking …

Student A	Student B
oil change?	oil filter wrench and a funnel
tighten screws?	Phillips screwdriver or slot screwdriver
loosen hexagon screws/bolts?	wrench, ratchet
shorten metal piece?	grip pliers (to fix the metal piece) and hacksaw
change tyres?	jack (to lift the car), wrench (to loosen nuts)
install car audio?	wire strippers
auto body repairs?	angle grinder
hammer out dents?	soft-faced hammer

C | Hand tools and power tools

R **1** Look at the website and read the Frequently Asked Questions (FAQs). Copy the chart and summarise the advantages and disadvantages of hand tools and power tools according to the FAQs.

Tools	Advantages	Disadvantages
Hand tools		
Power tools		

ww#.automotive_tools.uk

FAQs

How can tools be classified?
There are two main types of tools: power tools and hand tools.

What are the main advantages of hand tools?
Hand tools offer the benefit of precision. They are perfect for fine and detailed work such as repairing the electrical systems of cars or changing the fan belt on an engine. Also, hand tools are smaller and lighter than power tools, making them easier to transport.

What are the main disadvantages of hand tools?
The primary disadvantage of hand tools is the time that it takes to complete simple tasks. Changing the wheels using a tyre lever, for instance, takes much more time than using a tool powered by compressed air. Also, more skill is required to use hand tools than most power tools.

What are the main advantages of power tools?
Obviously power tools are a lot faster. Cutting huge metal bars with a handsaw is a slow and dangerous process. A metal-cutting machine offers the power needed to do the job quickly.

What are the main disadvantages of power tools?
One of the main disadvantages of power tools is that they are more complex than hand tools. The power connections, drive mechanisms, engines and gears wear out and can fail. Regular maintenance, such as sharpening, oiling, replacing parts is necessary. The costs of owning power tools continuously increase because of these maintenance and repair requirements. The risk of injury is also higher with power tools due to higher power and speeds.

R **2**
KMK I
A 4.2

Sie sollen für die Werkstatt Ihrer Firma einen neuen Schlagschrauber bestellen. Bei Ihrer Recherche stoßen Sie auf die britische Firma Milestone's Automotive Tools. Auf der Startseite gibt es eine kurze Vorstellung der Firma. Hören Sie sich diese an und beantworten Sie folgende Fragen.

1. Für welche Branchen verkauft die Firma Produkte?
2. Wodurch zeichnen sich die Werkzeuge aus?
3. Seit wann besteht die Firma?
4. Wo ist der Firmensitz?
5. Wie erhält man einen aktuellen Katalog?

P **3**
KMK I

Schreiben Sie eine E-Mail in Englisch an die Firma und bitten Sie freundlich um die Zusendung des neuen Katalogs innerhalb von zwei Wochen. Erkundigen Sie sich nach Lieferzeiten, Zahlungs- und Lieferbedingungen, Rabatten und Sonderangeboten.

WORD BANK

angle grinder – Winkelschleifer, Flex
centre lathe – Drehmaschine
column drill – Standbohrmaschine
fan belt – Keilriemen
funnel – Trichter
gear – Gang; Zahnrad
maintenance – Wartung
metal bar – Metallwerkstück
milling machine – Fräsmaschine
oil filter wrench – Ölfilterschlüssel
power drill – elektrische Bohrmaschine
wire stripper – Abisolierzange

→ **PHRASES:**
Written communication

B

NACH DIESEM MODUL:

- ✓ kann ich verschiedene Firmenabteilungen auf Englisch erklären,
- ✓ kann ich Arbeitsplätze und Tätigkeiten auf Englisch beschreiben,
- ✓ kann ich eine Firmenstruktur auf Englisch beschreiben.

Your company

Regardless of whether your company builds houses, repairs cars, or makes furniture, there is likely to come a time where you will be required to talk about yourself, your job and your company in English. Whether you are greeting an English-speaking visitor, giving a tour of the factory, or talking to suppliers from another country, it is important you have the skills and vocabulary to express yourself clearly.

1 **Work on your own. Look at the photos and brainstorm words that describe what is shown in each of them.**

I/P **2** **Compare your list of words with your partner and work together to describe the pictures. Use these words to help you:**

building site • cable • construction worker • grease • hard hat • car mechanic • overalls • ratchet • server • timber • tools • wires • work gloves

WATCH OUT!

She ~~wears~~ a hard hat. She is wearing a hard hat.

P **3** **With your partner, think of three questions you would like to ask each of the four young workers shown in the photos.**

A | The company and its departments

P **1** **Use the words and phrases in the table to explain what the departments of Krone Elektronik GmbH do.**

Example: Sales and Marketing advertises and sells the products.

Department		
Sales and Marketing	distribute	the products are not damaged.
Production Planning	*advertise and sell*	new employees.
Accounts	check	raw materials.
HR (Human Resources)	buy	*the products.*
Production	deal with	the computer network.
Logistics	recruit and train	the finished products.
Purchasing	maintain	customer problems.
Customer Service	handle	the company's finances.
R&D (Research and Development)	make	new products.
IT (Information Technology)	organize	production schedules.
Quality Control	develop	the products to the customers.

P **2** **Use the information in the table to form sentences using *responsible for / in charge of*.**

Example:
The Production department is responsible for making the products.
The Production Planning department is in charge of …

P **3** **Look at the departments again and write down three questions. Then close your book and ask your partner.**

Example:
A: What is the Quality Control department responsible for?
B: It is responsible for …
A: Which department is in charge of …?
B: The … department is in charge of …

R **4** KMK I A 1.2 **Der Ausbildungsleiter von Krone Elektronik, Herr Kirchner, zeigt Sean, einem Elektrotechnikstudenten aus Belfast, das Firmengelände und die einzelnen Abteilungen. Hören Sie sich das Gespräch an und beantworten Sie die folgenden Fragen auf Deutsch.**

1. Wann wurde Krone Elektronik GmbH gegründet, und von wem?
2. Wo hat die Firma ihren Hauptsitz?
3. Wie viele Produktionsanlagen hat Krone Elektronik insgesamt?
4. Wo wird nächstes Jahr eine neue Produktionsanlage eröffnet?
5. Wie viele Leute beschäftigt die Firma?
6. Woran arbeitet die Forschungs- und Entwicklungsabteilung gerade?
7. Welche drei Abteilungen befinden sich in dem Hauptgebäude?
8. Wo wird Sean zuerst arbeiten?
9. Wie viele Zylinderköpfe werden jährlich bei Krone Elektronik hergestellt?
10. Was passiert momentan in der IT-Abteilung?

WORD BANK

grease – Fett, Schmierfett
ratchet – Knarre
raw materials – Rohmaterialien
recruit (v) – anwerben
tool – Werkzeug

B | Tasks and responsibilities

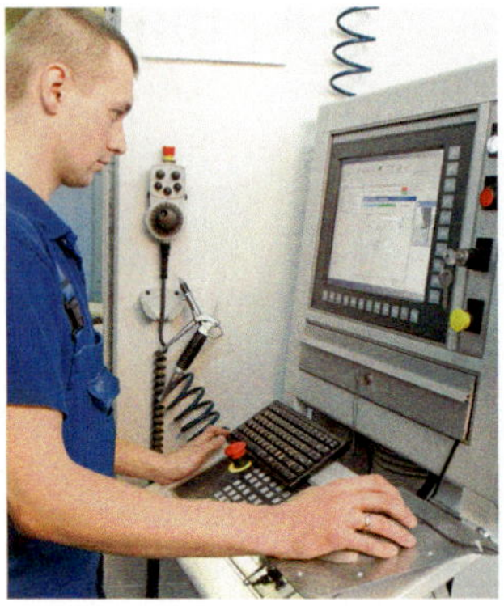

Industrial companies are made up of many different people, who all have a variety of tasks and responsibilities. DWG Tech GmbH in Stuttgart manufactures machine parts for the automotive and aviation industry. Below, Tim Schieber, a DWG Tech GmbH employee, talks about his job and daily activities on DWG Tech GmbH's English website.

R **1** **Read the text and answer the following questions in your own words.**

1. What do CNC machines do?
2. What problems can occur with CNC machines?

Hi, my name is Tim Schieber and I work as a CNC machine operator here at the DWG Tech headquarters in Stuttgart. I've been with the company since completing my apprenticeship three years ago and next summer I am going to work at the DWG Tech branch in Detroit.

In my opinion, what we do here is very important. CNC is all about precision, and it is therefore essential that I concentrate at all times. These machines drill, shape, cut and polish the parts for automobiles and aeroplanes. As a CNC machine operator I am responsible for one machine. As I said, this job is all about precision, and the machines are programmed specifically for each job. I do not do the programming myself, but I learnt programming during my apprenticeship and I sometimes have to intervene when something goes wrong – such as a vibrating workpiece or broken cutting tools.

The CNC process is almost entirely automatic. It is my job to set up the machine for the production process, which means loading the machine with the correct tools and workpieces. Once the programming for an operation is done, I enter the commands into the machine. To do this we use a computer network. The next step is to let the machine work its magic.

The main part of my job is to monitor the machine and spot problems. This means checking that the machine is running smoothly, measuring finished workpieces to check that the dimensions are correct, and sometimes replacing tools. It is also important that the machine is kept cool and lubricated with a special lubricating oil – but this is also done automatically. Of course, another of my tasks is to make sure the machine is kept clean, and that the working area is safe.

My working day depends on what shift I am on. We generally work either early shift or late shift. Early shift is from 6am until 2pm, and the guys on late shift begin at 2pm and clock out at 10pm. Sometimes, if we have a lot of work to do, there is also a night shift – but that doesn't happen too often. I actually prefer early shift. Even though I don't enjoy waking up at 4:45am, it is great to have the afternoon to do whatever I like – especially in the summer.

I really enjoy my job and I am proud to be working for an international company like DWG. My colleagues and I are always very busy and it can be fairly stressful. But I love working with the machines and it is interesting to see the many technological developments in the industry.

2 **Find the synonyms in the text for the following words:**

1. to finish	4. accuracy	7. to like
2. very important	5. to observe	8. workmates
3. cars	6. usually	9. quite

R **3** **Decide whether the statements are true or false. Give reasons for your decisions in German.**

1. Tim is currently doing an apprenticeship as a CNC machine operator.
2. Once the machine starts running, Tim's work is over.
3. One of Tim's tasks is making sure that the cutting tools are not broken.
4. Tim rarely works in shifts.
5. Accuracy is very important for CNC work.
6. The technology used in CNC machining has hardly changed over the years.

P **4** **Your boss has asked you to write a short text for your company's English website. Make sure you include some basic facts about your company and details about your daily tasks and responsibilities. Use some of the phrases in Tim's text, e.g. *I work as, I've been with the company since*, etc.**

→ **PHRASES:** Your company

P/I **5** **Check the Internet for a job in your industry that interests you. Make notes on the daily tasks and responsibilities.**

I **6** **Explain the job you researched to your partner, but don't mention the job title. Your partner must guess what the job is from the description you give. Then swap roles.**

C | Organisational structure

P/M KMK II **Sie sind Auszubildende(r) in der DWG Tech GmbH. Ihr Vorgesetzter benötigt eine Übersicht über die Ingenieursabteilung der DWG Tech GmbH Niederlassung in Detroit. Auf deren Firmen-Homepage finden Sie die folgenden Informationen zu den Mitarbeitern. Zeichnen Sie ein Organigramm nach der vorliegenden Beschreibung.**

Who's who in DWG Tech GmbH Detroit's engineering department?

DWG Tech GmbH Inc. employs around 60 people in its engineering department. The department is led by Chief Engineer Dr. Adam Foster, who is supported by his Personal Assistant Helen Green and DWG Tech's Assistant Chief Engineer Dr. Tim Bayliss. The engineering department is split into two core areas: automotive and aviation. Shift leaders Chris Falsone and Sandra Borowski report to Michael Kellermann, the foreman. They all belong to automotive which is run by experienced Chief Automotive Engineer, Dr. Torben Brinkema. DWG Tech's aviation department is in the capable hands of Dr. Judith Sommer, Chief Aviation Engineer, who has been with the company for over 15 years. Dr. Sommer is supported by foreman Frank Kapowski. Barry Richardson and Jens Schmelzer are the two shift leaders and work closely with Mr Kapowski.

WORD BANK

aviation – Luftfahrt
CNC (computerized numerical control) – CNC (computerisierte numerische Steuerung)
dimensions – Maße
drill (v) – bohren
lubricated – geölt
lubricating oil – Schmieröl
measure (v) – messen
polish (v) – schleifen, polieren
precision – Genauigkeit
shape (v) – fräsen, formen
vibrate (v) – vibrieren
workpiece – Werkstück

WORD BANK ⊙

braking system – Bremsanlage
cylinder head – Zylinderkopf
interface – Schnittstelle
manufacturing plant – Produktionsbetrieb
operating system – Betriebssystem

03

NACH DIESEM MODUL:

- ✓ kann ich mich zu den wichtigsten Prozessabläufen im Kfz-Rohbau auf Englisch äußern,
- ✓ weiß ich, wie einige Rohbauteile auf Englisch heißen,
- ✓ kann ich die wichtigsten Fügetechniken auf Englisch beschreiben.

The body shop

The body shop is the second stage in car production and follows immediately after the stamping shop. It is the place where all the individual stamped parts of a car are joined to form the car body. The car body is what surrounds the passengers and protects them. It is also called the body shell or passenger safety cell.

1 Talk to your partner and answer these questions.

1. What is happening in the picture?
2. What parts of the car body can you/can't you see in the picture?
3. How many car body parts do you know in English? Make a list.

2 Match the words to form body parts. Find the German translation of these words.

1. B	a. arch
2. boot	b. pan
3. door	c. lid
4. floor	d. panel
5. front	e. pillar
6. side	f. sill
7. wheel	g. wing

P **3** Use the terms for body parts in exercise 2 and form one sentence with each term.

A | What happens in the body shop?

1 Are the following statements true or false? What do you think?

1. The terms 'body shell' and 'passenger safety cell' are both used as synonyms for 'car body'.
2. Each vehicle is made from approximately 400 metal parts.
3. The body parts are welded in one process.
4. There is only one joining technology.
5. The upper body is welded together first.
6. The doors, the bonnet and the boot lid are all called 'hang-on parts'.
7. Before the body leaves the body shop it is inspected.
8. When the body leaves the body shop it goes to the assembly shop.

R **2** Now read the text below and check your answers in exercise 1.

Building the 'body in white'

In the body shop all the stamped **sheet metal parts** are joined together to form the raw metal shell of the car, the so-called 'body in white'. To do this, several different joining technologies are used. When you look at a car, most of what you can see is part of the body. As you can imagine, you need a lot of parts to build a car. A standard saloon car, for example, can have around 400 parts. First these parts are joined together in separate robotic stations to form **sub-assemblies** such as the **front unit**, rear unit, centre unit or side panel. The robots then position these sub-assemblies very exactly on the **jig pallet** which moves along on the **conveyor** to the different welding stations. For example, the front unit and rear unit are welded to the centre unit to create the **underbody**. Then the side panels, roof and other parts of the upper body are added to form the complete metal frame of the car. An online monitoring system checks all the joining technology processes to ensure quality is always high. That means that every single joint or weld is continuously checked and regulated. Finally, the hang-on parts, that is the bonnet, doors, wings, boot lid and tank lid are screwed onto the **frame** in special geometry welding stations. At this stage the metal passenger safety cell, the 'body in white' is finished and receives its final inspection. The body is checked for smoothness and the measurements of the car bodies are checked with sensors to make sure there are no deviations. The whole body has to have been produced according to the specifications. Only perfect bodies go on to the paint shop.

3 Match the translations below to the words in bold in the text above.

Rahmen • Unterbau • Vorbau • (Unter-)Baugruppen • Bauteilträgerwagen • Fließband • Blechteile

4 Some terms are different in British and American English. Match the terms.

BE:	1. bonnet • 2. boot • 3. wing • 4. windscreen • 5. saloon • 6. sill
AE:	a. sedan • b. rocker • c. windshield • d. fender • e. hood • f. trunk

5 UNDERSTANDING TECHNICAL ENGLISH

Make sentences with these words to show their technical and everyday meaning.

1. boot 2. hood 3. trunk 4. saloon 5. body 6. wing

WORD BANK

assembly shop – Montagehalle
body in white – Rohkarosserie
body shell – Rohbaukarosserie
body shop – Karosseriewerkstatt, Rohbau
hang-on part – Aufhängeteil
joint – Verbindung, Fuge
measurement – Abmessung, Maß
metal shell – Metallgehäuse
rear unit – Heckteil
saloon – Limousine
sill – Schweller
weld (v)/(n) – schweißen / Schweißnaht

B | Joining technologies in the body shop

1 Match the joining technology on the left to its definition on the right.

1. adhesive bonding 2. brazing 3. rivet joining 4. spot welding 5. laser welding	a. Joining two pieces of metal with a heated metallic filler. When the filler metal cools, it forms a strong joint. b. A fusion process used to join the edges of similar materials of different thicknesses using a laser. c. A technique used to bond sheet metals no thicker than 3 mm. The metal pieces are clamped together with two electrodes and a strong electric current is run through them. d. A mechanical joining method often used to clamp materials together that cannot physically be welded together. e. A method of bonding materials together using glue.

2 Match the numbers of the sub-assemblies in the picture with the English terms below.

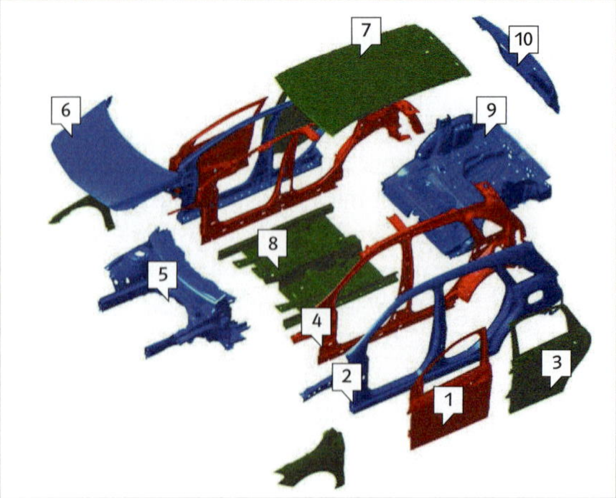

a. roof	f. sidewall outer
b. sidewall inner	g. bonnet/hood
c. floor unit	h. tailgate
d. front door	i. rear door
e. front unit	j. rear unit

P **3** You will hear an interview about joining technologies in the body shop. Before you listen, rearrange the words below to form questions.

1. joining • common • is • most • technology • today • used • what
2. there • methods • many • joining • how • different • are
3. a • both • combination • do • of • use • when • you
4. assured • body • how • in • is • quality • shop • the
5. advantage • is • laser • of • the • welding • what

4 Take turns to ask and answer the questions you formed in exercise 3.

R **5** Listen to the interview. Does Bernd give the same answers as you did?
◎ A 4.3

R **6** Listen again and finish the following sentences.
◎ A 4.3

1. The most common joining method we use today …
2. Laser welding is much …
3. Aluminium and steel cannot …
4. The joints also make the car body …
5. When we need a very stiff body, we often use …

C | Materials to build the body in white

R **1** Read the audioscript on pages 119 and 120 and find out what Bernd has to say about materials used in the body shop. Make a list of the materials he mentions.

R **2** Look at the audioscript again and answer the following questions in complete sentences.

1. Why does the material have to be stiff?
2. Name three different materials that are used to make the car body.
3. What material is used for the B-pillar and why is it used?
4. Why is aluminium used?
5. Why is steel still used to make hang-on parts?
6. What do they hope to use in future for parts like the longitudinal member?

3 Look at the two pie charts and compare the materials used to build the old and new model. Write six sentences comparing the differences.

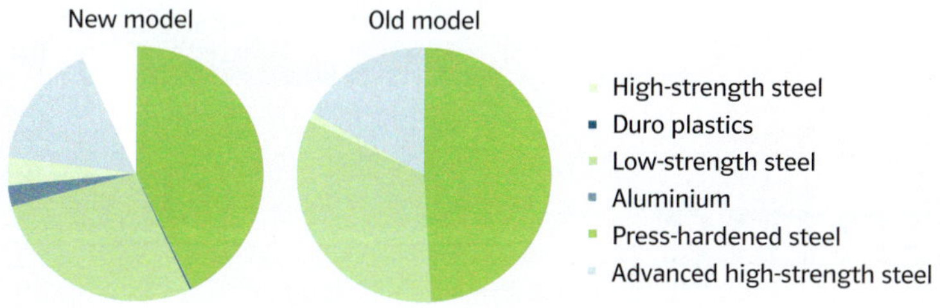

4 Which materials would you use for which parts of the body in white? Make a list.

M **5** Ihr Abteilungsleiter bittet Sie, den folgenden Text ins Deutsche zu übertragen.
KMK II

The body shop and the environment

There are not many environmental issues that have to be considered in the body shop because there is almost no waste in the body shop. Not like in the press shop for instance. You have a lot of waste material there because there is always some sheet metal left over when the shapes are cut out. But we do have one problem in the body shop: all the sheet metal parts are coated with oil to protect them against corrosion during the production process. We have to make sure no toxic fumes are produced during the welding process so that the technicians don't inhale any poisonous gases and get sick.
Apart from that all the other enviromental issues in the body shop are really more to do with reducing energy. They are linked to the energy used to produce a car (e.g. the robots). Large efforts are being made to make sure that robots only use energy when they are actually in operation. The goal is for them to use no energy when they are idle, like when they are waiting for the next vehicle. It sounds simple – but it's really quite demanding for the control unit.
Another goal is to reduce the energy consumption of the joining technologies themselves. We have come a long way there, especially with the new innovative laser welding technology. Laser normally uses a lot of energy but new laser sources only need 30 % of the energy the old laser sources used.

WORD BANK

adhesive bonding – Klebeverbindung
clamp (v) – befestigen, einspannen
current – Strom
longitudinal member – Längsträger
stiff – steif, fest
thickness – Stärke, Dicke
toxic fume – giftiges Gas

WORD BANK ◎

cold-/hot-formed – kalt-/warmumformt
crimp (v) – quetschen, kräuseln
fibre reinforced – faserverstärkt
fuse (v) – verschmelzen
high-strength – hochfest, hochwiderstandsfähig
melt (v) – schmelzen
rivet gun – Nietpistole
self-piercing rivet – selbststanzende Niete

C

NACH DIESEM MODUL:

- ✓ kenne ich die wichtigsten Redewendungen um ein Telefongespräch zu führen,
- ✓ kann ich Maße, Namen und Adressen am Telefon auf Englisch weitergeben.

Telephoning

When you are on the job, chances are you will have your mobile phone with you – it does not matter what line of work you are in. Whether you are repairing a server, fixing a leak, or installing a telephone line, your phone can ring at any time. It could be a customer with a complaint, you may have to ring a supplier or check the status of an order. And if you have to talk in English, you need to have the necessary skills to understand and be understood, no matter what the situation.

P **1** **Look at the photos. What might the people be talking about on the telephone? Compare your answers with a partner. Use these words to help you:**

materials • problem • delay • support • help • breakdown • repair • solution • angry • confused • crash • computer network room

2 **Match the German words below with the English equivalents.**

1. Handy	a. message
2. Anrufbeantworter	b. hands-free device
3. Festnetz	c. extension number
4. Nachricht	d. mobile/cell phone
5. Durchwahl	e. answering machine
6. Freisprechanlage	f. area code
7. Vorwahl	g. country code
8. Ländervorwahl	h. landline

A | Making and receiving a phone call

R **1** **Which of the following phrases are used for making calls and which ones for receiving calls? Make a list.**

1. Could you put me through to Mr Warren, please?
2. I'm sorry, but Mr/Ms … is in a meeting at the moment.
3. How can I help you?
4. Thank you. I'll ring back later.
5. Can I ask what your call is about?
6. Would you like to leave a message?
7. Thank you very much for your help.
8. My name is Barry Potter and I'm calling from …
9. May I take your name, please?
10. I'd like to speak to somebody in the production department, please.
11. Hold the line, please. I'll put you through.
12. Could you tell him/her that …?

M **2** **Read the telephone conversation below and find the English equivalents for the following German expressions.**

→ **PHRASES:** Telephoning

1. Haben Sie etwas zum Schreiben?
2. Soll ich Ihnen seine Handynummer geben?
3. Er ist leider zurzeit nicht im Büro.
4. Ich hätte gerne mit … gesprochen.
5. Einen Moment, bitte.
6. Wie kann ich Ihnen helfen?

Christoph:	Smart EDV, Christoph Fink.
Caller:	Er, yes, hello. Do you speak English?
Christoph:	Certainly. How can I help you?
Caller:	This is the Smart IT company in Kiel, isn't it?
Christoph:	Yes, that's correct.
Caller:	Great. My name is Erik Pedersen and I'm calling from Danske Onlinebank Copenhagen. I'd like to speak to Mr Gebhardt, please.
Christoph:	Sorry, but Mr Gebhardt is not in the office at the moment. Can I take a message, or would you like his mobile number?
Caller:	Can you give me his mobile number, please?
Christoph:	Ok, do you have a pen?
Caller:	Just a second, please … ok, go ahead.
Christoph:	The number is 0049-170 557624.
Caller:	Sorry, I didn't catch that. Could you speak up a bit?
Christoph:	Yes, of course. The number is 0049-170 557624.
Caller:	Thank you very much.
Christoph:	You're welcome. Maybe you can give me your phone number.
Caller:	Certainly. It's 0045, that's the country code for Denmark, 32, the area code for Copenhagen, and then 54 58 11. Oh, and my extension number is 944.
Christoph:	0045-32 54 58 11, extension 944. Thank you very much. Goodbye.
Caller:	Goodbye.

WORD BANK

breakdown – Störung, Ausfall
crash – Absturz
leak – Leck

INTERCULTURAL BOX

Intercultural awareness – Telephoning in English-speaking countries

When talking on the phone in English it is important that you use suitable and polite phrases. For example, if you know the person you are calling, it is common to make a little small talk before getting down to business. One thing to remember: Although it is common for English people to quickly switch to first names, it is still important to remain polite and remember that you are talking to a customer or a colleague, and not your best friend.
Example: *How are you? I haven't spoken to you for a while.*
If you have to give bad news or talk about a problem, then you should begin with *I'm afraid …* or *Unfortunately …* .
Example: *I'm afraid we cannot install your wireless internet connection this week.*
A request often begins with *Could you possibly …* or *I would be grateful if you would / could …* .
Example: *I would be grateful if you would call me back later.*
When someone thanks you for your help, or for a job well done, you can simply say *You're welcome.* And be careful! Whereas in German, just saying *yes* or *no* may be fine, this is considered to be quite impolite in England. Instead, just say *Yes, I think so*, or *No, I'm afraid not*. If you call somebody, you should give your name, where you are calling from, and say who you would like to speak to.
Example: *Hello, my name is Felix Hirschbach, calling from Crown Electronics. Could I speak to Ms Cameron, please?*

P **3** **How would you say the following things more politely?**

→ **PHRASES:** Telephoning

1. Eh? Say it again!
2. Give me your number!
3. What do you want?
4. What's your name?
5. I can't hear you!
6. What?
7. He's not here!
8. Give me Ms Bauer!

I **4** KMK I **Sie arbeiten im International Calls Service Center von Smart EDV in Kiel und erhalten einen Anruf. Führen Sie das folgende Telefongespräch mit einem Partner auf Englisch.**

Smart EDV GmbH, Kiel.	Stefanie Berg, dänische Kundin
Nehmen Sie den Anruf entgegen. →	Nennen Sie Ihren Namen. Sie möchten mit Herrn Gebhardt sprechen.
Sagen Sie, dass Sie die Anruferin schlecht verstehen und bitten Sie sie, etwas lauter zu sprechen.	Sprechen Sie etwas lauter und fragen Sie, ob Sie jetzt besser zu verstehen sind.
Bejahen Sie und fragen Sie, worum es geht.	Erklären Sie, dass die Firma Smart EDV gestern in Ihrer Firma neue Hardware und Software installiert hat, und dass heute Morgen der Server abgestürzt ist. Es ist dringend!
Sagen Sie höflich, dass Herr Gebhardt in einer Sitzung ist und sich später melden wird. Fragen Sie, ob Herr Gebhardt die Nummer der Anruferin hat.	Bejahen Sie. Bedanken Sie sich und beenden Sie das Gespräch.
Bedanken Sie sich auch und verabschieden Sie sich.	

B | Giving information over the phone

Making a quick call is often the fastest and easiest way of passing on valuable information, such as measurements or figures. However, misunderstandings can happen very quickly, so it is important to make sure any information is delivered clearly and correctly.

1 Match the English word to the German translation.

1. height	a. Breite
2. length	b. Gewicht
3. width	c. Höhe
4. weight	d. Tiefe
5. depth	e. Länge

INFO BOX: Saying weights and measurements

nine **point** five millimetres (9.5 mm)	100 **by** 30 **by** 15 (100 x 30 x 15)
thirty one **square** metres (31 m²)	twenty **cubic** metres (20 m³)
one **point** two five litre engine (1.25 l)	6000 revolutions **per** minute (6000 rpm)
170 kilometers **per** hour (170 kph)	two hundred **and** twenty volts (220 V)

2 What are the adjectives for the nouns in exercise 1?

I **3** Sit back to back with a partner. Take turns in giving each other different information over the phone. Partner A uses role card A, Partner B uses role card B (see Appendix).

I **4** Sit back to back and spell your name, the name and address of your company, and your email address. Use the table below to help you.

Symbol	Name	Example
@	at	info@
B / b	capital letter / lower case	NYC / asap
-	hyphen / dash	t-online
ö	o-umlaut / oe / o with two dots	Schönberg
:	colon	http:
/	slash / stroke / forward slash	org/news
\	backslash	\docs
.	dot	.de
_	underscore	customer_info

R **5** KMK I ⦾ A 1.3 Alex Bauder works for Fertighaus Henke in Magdeburg and is called by Mr. Reid, an English customer who lives in Germany and who is currently building a house near Magdeburg. Listen to the conversation and answer the following questions in German.

1. Who is Mr Deschler?
2. Why can't Mr Reid speak to him personally?
3. Why does Mr Reid want to speak to Mr Deschler?
4. When will Fertighaus Henke begin building Mr Reid's cellar?
5. Why does the extractor fan need to be moved to the left?
6. Which changes is Mr Deschler not responsible for?
7. What should Mr Deschler do next?

R **6** ⦾ A 1.3 Now listen to the conversation again and make notes of the changes Mr Reid wants, where he wants them, and details about the measurements, where given.

WORD BANK

cubic metre – Kubikmeter
revolutions per minute – Drehzahl
socket – Steckdose
square metre – Quadratmeter

WORD BANK ⦾

construction manager – Bauleiter
diameter – Durchmesser
extractor fan – Dunstabzugshaube
foundation pit – Baugrube
hob – Kochfeld

04

NACH DIESEM MODUL:

- ✓ kann ich den Ablauf der Oberflächenbehandlung in der Kfz-Herstellung auf Englisch erklären,
- ✓ weiß ich, wie ich auf Englisch Fragen zu Prozessabläufen formulieren kann,
- ✓ kann ich die wichtigsten Fehler beim Lackieren beschreiben.

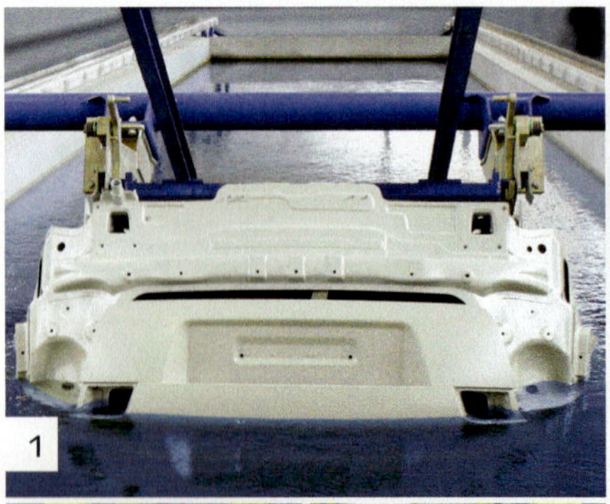
1

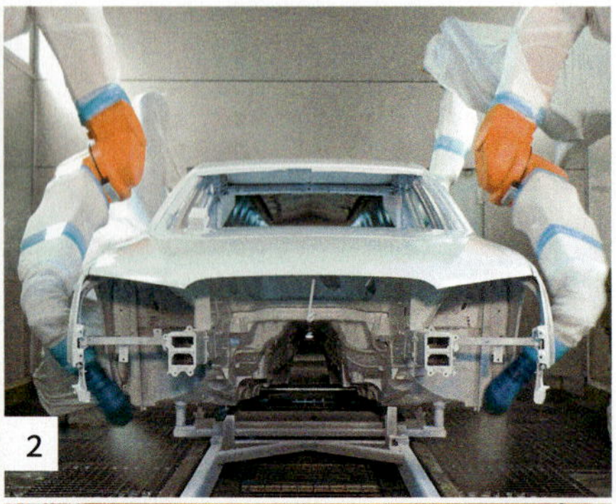
2

3

4

The paint shop

The paint shop is an important part of a car production factory (also often called 'plant'). The paint shop is where the body of the vehicle is painted and gets its 'coat'. The painting stage is the third stage in car production and follows directly after all the individual parts have been joined together in the body shop.

1 Answer the following questions.

1. What do you think is happening in the pictures?
2. Why do you think the car body is dipped into a bath?
3. What are the main reasons for painting the car body?
4. How do you think the car is dried?

2 **UNDERSTANDING TECHNICAL ENGLISH**

1. The words 'body', 'coat' and 'shop' have specific meanings in car production. Which expression in each line is **not** used to talk about car production?

   a. car body • body repair shop • body shell • body language • body shop
   b. base coat • clear coat • e-coat • overcoat • top coat • undercoat
   c. assembly shop • body shop • sweet shop • press shop • stamping shop • paint shop

2. Form a sentence with each of the above expressions to illustrate its meaning. Use a dictionary if necessary.

A | Stages in painting a car

R **1** Read the text about the different stages in painting a car and check your answers to the questions in exercise 1 on page 32.

How to paint a car

Painting the body not only gives the car its shiny bright colour; more importantly it protects it against rust, corrosion, stone damage and the effects of extreme weather conditions and UV rays.

Before the body can be painted, any remaining dirt or grease must be removed. Only when the car body is spotless can it receive its initial protection against corrosion. The body travels on a conveyor system and enters a large tank to receive its first application: the phosphate coat. This coat is not just a protective coat, it also forms an ideal base for the second layer of paint.

Next, the body is dipped into a cathodic bath to receive its second protective undercoat: the e-coat. It is called 'e-coat' because it is applied with the use of electrical charges. Immersing the body into this big tank ensures that even the smallest cavity in the body shell is sealed. The body shell then leaves the cathodic bath and is dried in an oven that has a temperature of 190°C to harden the paint. This drying process is called 'curing'. Then the underbody is treated and the seams are sealed.

After the body shell has received this corrosion resistant, protective coat of paint, it is checked and if necessary sanded to remove any surface dirt. It is cleaned again so that it is ready to continue on to the next stage. This is when the primer, base coat and clear coat are applied. Absolute cleanliness is vital here.

M **2** Your boss needs an information sheet with the main steps of the painting process for new colleagues. Write the information sheet in German based on the text above.

3 Match the words to form common word combinations used in the paint shop.

1. apply	a. based
2. base	b. bath
3. cathodic	c. coat
4. clean room	d. friendly
5. clear	e. paint
6. corrosion	f. performance
7. dust-free	g. resistant
8. environmentally	h. technology
9. high	i. zone
10. water	j. coat

4 List the word combinations in exercise 3 on a separate sheet of paper and add the German translations. If necessary use a dictionary for help.

P **5** Check the Internet to find out which treatments are carried out on the underbody to make the car quieter and to protect it against corrosion. Present your findings in class and include at least three of the expressions from exercise 3.

WORD BANK

application – *(hier:)* Anstrich
apply (v) – *(hier:)* auftragen
base coat – Basislack
body shell – Rohbaukarosserie
body shop – Karosseriewerkstatt, Rohbau
cavity – Hohlraum
clear coat – Klarlack
coat – Beschichtung, Überzug
corrosion resistant – korrosionsbeständig
curing – Aushärten
dip (v) – tauchen
grease – Fett
immerse (v) – eintauchen
paint shop – Lackiererei
primer – Grundierung
sand (v) – absanden, abschmirgeln
seal (v) – versiegeln
treatment – *(hier:)* Aufbereitungsverfahren
underbody – Unterbau
undercoat – Grundierung

B | A visit to the paint shop

1 Michael and Anna have been invited to see the last stages of the painting process of a large car plant. They have been asked to prepare some questions. Look at the questions in the box and explain the difference between them. Which question form is used when you are not really interested in who exactly performs the action?

Active	Passive
How do you obtain a dust-free zone? How do you clean the cars? How long do you take to paint a car?	How is a dust-free zone obtained? How are the cars cleaned? How long does it take to paint a car?

2 Change the active questions below into passive questions. Look at page 139 for help.

1. Why do you need a dust-free zone in the paint shop?
2. What kind of paints do you use?
3. How do you paint the car bodies?
4. When do you check the quality of the painted car body?

R **3** ◎A 4.4 Listen to Mr White talking to Michael and Anna about the coating process in the paint shop. Does he answer all of the above questions?

R **4** ◎A 4.4 Listen again and note down the answers Mr White gives.

P **5** Use your notes and the words below to prepare a flowchart of the different stages of the coating process.

first of all • secondly • next • after that • then • in the following stage • finally

P **6** KMK II Michael ist gebeten worden, eine kurze Präsentation über den Ablauf der Oberflächenbehandlung in der Lackiererei zu geben. Er schreibt eine höfliche E-Mail an Herrn White, um sich für die Führung zu bedanken und ihn gleichzeitig um Hilfe bei der Erstellung der Präsentation zu bitten. Schreiben Sie Michaels E-Mail und berücksichtigen Sie dabei die folgenden Punkte.

→ PHRASES: Written communication

- Bedanken Sie sich dafür, dass Herr White sich die Zeit nahm, Ihnen und Anna die Oberflächenbehandlung zu zeigen und zu erklären.
- Sagen Sie, dass Sie das Tauchbad besonders interessant fanden.
- Erklären Sie, dass Anna alles so interessant fand, dass sie gerne später in einer Lackiererei arbeiten würde.
- Teilen Sie Herrn White mit, dass man Sie gebeten hat, am 25. August eine kurze Präsentation über die Oberflächenbehandlung in der Lackiererei zu halten.
- Fragen Sie, ob er Ihnen bei der Folienerstellung behilflich sein könnte und ob es möglich wäre, ein paar Bilder für die Präsentation zu bekommen. Ideal wäre ein Bild von jedem Abschnitt des gesamten Prozesses.

C | Quality problems in the paint shop

1 Match the defects to their definitions.

1. blemish	a. Here the surface looks like the outside of an orange.
2. blister	b. You can see very small bubbles on the surface.
3. chip	c. You can see an inward deformation in the metal.
4. crater	d. This looks like an extra drop of paint on the coat.
5. dent	e. These appear when something hard and sharp goes across the surface.
6. drip	f. It looks like a long downward line of paint and often ends with a droplet.
7. orange peel	g. This is sometimes also called 'pin hole'.
8. run	h. A small part of the coat of paint has been broken away.
9. rust	i. This is another word for 'defect'.
10. scratch	j. This is a type of corrosion. It is often brown.

2 Copy the following table and fill in the defects in exercise 1.

Defects caused from the outside	Defects caused during painting

M **3** Put the list of defects in exercise 2 into German.

I/P **4** Which three painting defects are shown in the pictures? Compare your answers with a partner and discuss what you think may cause these defects.

orange peel • chip • blister • drip • run

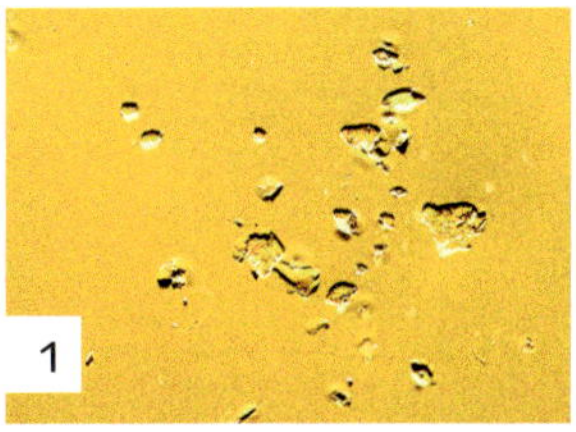

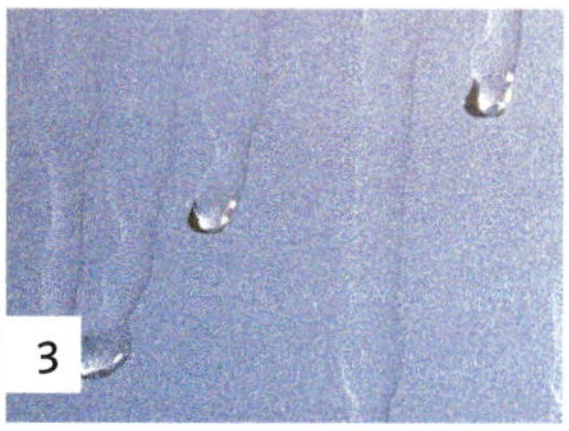

P **5** Form three groups. Each group takes a different defect. Search the Internet to find out some of the causes and how you can repair the defect. Make notes and then present your findings to the class.

6 **UNDERSTANDING TECHNICAL ENGLISH**
The following words are often used in a paint shop. Use a dictionary and find their general and technical meanings.

1. blister 2. chip 3. crater 4. run 5. scratch

P **7** **SAFETY FIRST** ⚠
Make a list of safety tips for working in a paint shop. Compare and discuss your lists in class.

WORD BANK

chip – Abplatzung
dent – Delle, Beule
droplet – Tröpfchen
pin hole – Nadelloch

WORD BANK ◎

airlock chamber – Luftschleuse
assembly – Montage
blemish – Makel
impurity – Verunreinigung
inject (v) – einspritzen
nozzle – Spritzdüse
solvent – Lösungsmittel

D

NACH DIESEM MODUL:

- ✓ kann ich auf Englisch kurze Anfragen, Angebote und Bestellungen als Brief, E-Mail oder Fax erstellen,
- ✓ kann ich englische Briefe, E-Mails und Faxe beantworten.

1

2

3

4

Written communication

Written communication such as emails, faxes and business letters play an essential role in business, both in the office and the workshop. For any company to run efficiently, it is necessary to have an effective communication system – making sure that enquiries are answered promptly, offers are made on time, and orders are placed and processed correctly. The standard business letter has been replaced in recent years by the faster email, and to a lesser extent, the fax. In all areas of business today it is common to have dealings with countries across the globe, meaning you need to be able to understand written communication in English.

P **1** **Look at the pictures above and describe them in your own words.**

P **2** **What are the advantages and disadvantages of the forms of written communication shown in the pictures? Create a list. Use these words to help you:**

confidential • instant • easy • time-consuming • reliable • signal • misunderstanding • slow • personal • formal • informal

3 **Compare your lists with a partner and come up with a final list for each form of communication. Share your final list with the class.**

A | Enquiries

Business transactions generally begin with an enquiry. In the example below, a car garage is looking for a new supplier of tools.

M **1** **Stephen Bains from City Garage in Wolverhampton, England is interested in buying some tools from Hauch Werkzeuge KG in Gelsenkirchen, a toolmaking company specializing in tools for car mechanics. Read Stephen's enquiry and sum up the main information in German.**

Dear Sir/Madam,	a. appropriate salutation
Automotive mechanic tool supplier	b. subject line
We saw your advertisement in this month's edition of *Under the Bonnet* magazine regarding your selection of tools for car mechanics.	c. source of address
We are a family-run car mechanic company in Wolverhampton, England, with over 20 employees.	d. introduction of your firm
We have recently expanded and are looking for a new supplier of quality tools for our new workshops. Therefore we would ask you to send us a copy of your latest catalogue and price list.	e. reason for enquiry f. request to send catalogue and price list
Please also inform us about your export prices and possible discounts as well as terms of payment and delivery and delivery times.	g. request for information on prices and discounts and terms of payment and delivery
As we would like to test your tools, we would appreciate it if you could send us some samples.	h. further requests
Thank you for your attention to this enquiry and we look forward to hearing from you.	i. closing phrase
Yours sincerely,	j. complimentary close

R **2** **Match these parts of an enquiry with the descriptions (a.–j.) above.**

1. Best wishes.
2. Your company was recommended to me by a colleague.
3. We are a large production company, specialising in machine components.
4. Please let us have a detailed cost estimate for configuring our IT network.
5. We hope to hear from you soon.
6. What are your terms of payment for regular customers?
7. As we are convinced that your products will sell well on the American market …
8. A visit by your representative would be greatly appreciated.
9. Please let us know whether you can supply from stock.
10. Machine component supplier

WORD BANK

bonnet – Motorhaube
car mechanic – KFZ-Mechaniker(in)
toolmaking company – Werkzeugmacher
workshop – Werkstatt

B | Offers

P **1** **Read the offer email below and complete it using the words from the box.**

bank transfer • discount • interest • business • working days • sample • attachment • delivery period • stock • orders

From: mailto:t.wolf@hauch...de **Sent:** 201_-05-30 14:52
To: mailto:info@citygarage.co...uk
Cc: mailto:k.mueller@hauch...de
Subject: Your enquiry

Attachments Hauptkatalog_2012.pdf

Dear Mr Bains

Thank you for your enquiry and your **1** in our products.

We are sending you our latest brochure and price list as an **2**.
For **3** of 200 units or more we are able to grant a quantity **4** of five per cent.
For large orders (over 250 units) we require a **5** of 10–14 **6**.
Smaller orders can be delivered from **7** and within 3–5 working days.

Our usual terms of payment are by **8** to our account with the GDB Bank in Gelsenkirchen.

One **9** case of ratchets, sockets, wrenches and screwdrivers has been dispatched to you this morning and should reach you in the next few days.

Thank you again for your enquiry. We look forward to establishing regular **10** with you.

Yours sincerely

Tim Wolf

P **2** **Use the following notes to write an enquiry and an offer in English.**

→ **PHRASES:** Written communication

Enquiry
(You work at Construct World.)

- large DIY store in Poland (Construct World), looking to buy quality German-made tools
- saw your company (Hauch Werkzeuge KG) at a trade show in Hannover
- would like information about delivery times, discounts, copy of catalogue
- would like some samples to inspect
- polite ending

Angebot
(Sie arbeiten bei Hauch Werkzeuge.)

- Bedanken Sie sich für die Anfrage.
- Gerne schicken Sie Ihren aktuellen Prospekt inklusive Preisliste zu.
- Lieferzeit (nach Polen):
  große Bestellungen (über 500 Stück): 5–10 Werktage,
  kleine Bestellungen: 2–4 Werktage
- Preisnachlass ab Bestellwert von 2.500 € möglich (5 %)
- Sie würden sich freuen, mit Construct World zusammenarbeiten zu dürfen.
- höflicher Schluss

C | Orders

1 **Read the order fax below and complete it using the prepositions in the box below.**

for (2x) • by • of • to (3x) • with

TELEFAX Message

City Garage
17 Gorsebrook Rd.
Wolverhampton WV6 5PI
Tel.: 0044 (0)1902 554997... • Fax: 0044 (0)1902 554998...

To: Hauch Werkzeuge KG
Schmiedstrasse 44
45879 Gelsenkirchen
Germany

Attention: Mr Tim Wolf

Fax: 0049 209 721965...

From: Stephen Bains

Subject: Order

Dear Mr Wolf

Thank you very much **1** your email **2** 30 May.

Thank you also for sending us your catalogue and price list.
We were very impressed **3** your range of tools for car mechanics and would like to place a trial order **4** 16 socket sets, 16 screwdriver sets, 30 adjustable wrenches and 16 torque wrenches.

Please confirm this order indicating the bank account **5** which you wish to have the sum in question transferred.

Please deliver the tools **6** our workshop at 17 Gorsebrook Rd., Wolverhampton, WV6 5PI.

We look forward **7** receiving the consignment soon and to placing further orders **8** you.

Best regards

S.Bains

Stephen Bains

M **2** KMK II **Sie arbeiten bei Hauch Werkzeuge KG in Gelsenkirchen. Ihr Chef bittet Sie, das obige Fax für ihn auf Deutsch zusammenzufassen, damit er über den Auftrag informiert ist.**

WORD BANK

ratchet – Knarre
screwdriver – Schraubenzieher / -dreher
socket – Steckdose, Stecknuss
torque wrench – Drehmoment-schlüssel
(adjustable) wrench – (verstellbarer) Schraubenschlüssel

05

NACH DIESEM MODUL:

- √ kann ich mich auf Englisch zu Fahrwerk und Antrieb äußern,
- √ kann ich die Funktion von Kfz-Federungssystemen beschreiben.

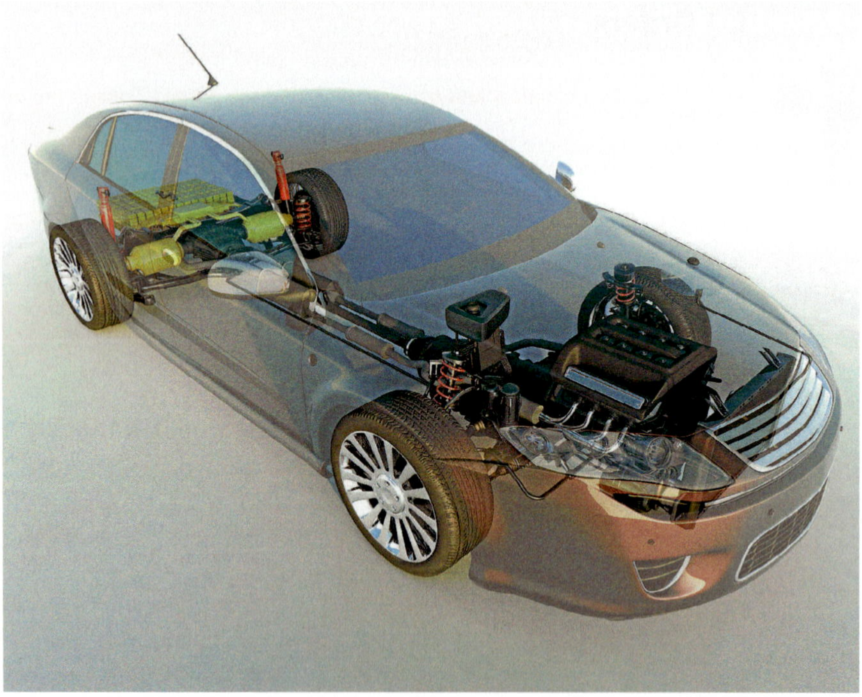

Chassis and drivetrain

The term 'chassis' is used in different ways but technically the chassis is an ultra-strong frame underneath the vehicle. It is usually made of high-strength steel alloys and it is what keeps the body, suspension and wheels together. There are basically two main types of chassis constuction: body-on-frame and unit-body construction. The drivetrain is also known as the powertrain. As the term suggests, it is everything needed to send the power from the engine to get the wheels moving so you can drive the vehicle. In other words it connects the transmission to the drive axles of the car. The main types of drivetrain are RWD (rear wheel drive) and FWD (front wheel drive).

1 Describe what you can see in the picture above.

2 Say what you know about body-on-frame chassis and unit-body chassis.

3 What are the differences between a rear wheel drive car and a front wheel drive car? Make a list.

4 Compare your list with two other pairs and take notes of additional differences.

5 Think of three FWD vehicles and three RWD vehicles you would like to drive and present them to the class.

A | Chassis construction

P/I **1** Imagine you were going to interview an engineer on the chassis of a vehicle. Make a list of questions you would ask him or her. Then role-play the interview.

→ **PHRASES:**
Meeting people, Your company

R **2** Match the questions to the answers below .

Questions:

1. Can you tell me what exactly the chassis of a car is? I've read so many different things in books and magazines.
2. The body is almost everything you can see when you look at the car, isn't it?
3. So that means the axles and wheels?
4. But not for passenger cars?
5. What is the suspension system?
6. Is the unit-body car normally stiffer than the body-on-frame vehicle?
7. What else is better?
8. OK – now I understand, because that means it uses less fuel, right?

Answers:

a. Yes, and that's why we now include the body and hang-on parts when we talk about the chassis!
b. That's because today, the body itself is built to be much stronger. It's in itself a cocoon, a safety cell. When cars were first built the body was built by coach makers and was not strong enough to have wheels attached to it.
c. That's basically everything that helps you steer the car and make the ride smooth. So to even out the bumps in the road, etc. The shock absorbers form a large part of the suspension system.
d. Yes – it's confusing, isn't it? That's because technology has moved on since the first chassis was developed. But let's start at the beginning: in the traditional sense the chassis is basically everything except the body.
e. Yes, the front axle, rear axle and wheels form part of the chassis. Also the suspension system and the steering system. We call that the body-on-frame construction and it is still used for trucks and SUVs today.
f. Exactly, the chassis is everything underneath the body. It's the frame of the bottom of the car and has all the parts that enable the car to work attached to it.
g. No, in passenger cars we normally talk about unit-body construction. Today we have different materials and alloys so the body of the car is much stronger. A great deal has been done to strengthen the body.
h. Exactly.
i. Well, it's also a lot lighter!

R **3** Now listen to the recording and check your answers in exercise 2.
◎ A 4.5

WORD BANK

body – *(hier:)* Karosserie
chassis – Fahrgestell
drive axle – Antriebsachse
drivetrain / powertrain – Antrieb
FWD (front wheel drive) – Vorderradantrieb
unit-body construction – Schalenbauweise, Monocoque
RWD (rear wheel drive) – Hinterradantrieb
shock absorber – Stoßdämpfer
steer (v) – lenken, steuern
suspension – Federung, Aufhängung
transmission – Getriebe

B | Suspension systems

R **1** **Read the extract from a website and answer the questions.**

1. What are the main functions of a suspension system?
2. What happens when a vehicle dives?
3. What could happen if the car is standing still and you suddenly accelerate?
4. What do each of the pronouns 'it' in the text below refer to?

ww#.suspension_info.uk

A suspension system ensures good roadholding, safe handling and easy cornering. It is what helps the tyres of a vehicle grip the surface of a road so that a driver can steer it easily and accurately – even when going around tight bends, driving over bumpy surfaces, braking or accelerating. It helps give the driver a comfortable ride by dampening the up and down movement of the wheels so they stay on the surface of the road. That makes the ride also safer. A good suspension system can prevent the car from diving or squatting. A vehicle can dive during heavy or sudden braking. That means it goes down at the front and up at the rear. Sudden, hard acceleration can cause a vehicle to squat. Then the car rises at the front and drops at the rear. In severe cases the front or rear wheels can even lose contact with the surface of the road making it impossible to steer the vehicle.

2 **Complete the following text using the words in the box.**

absorbing • axles • bounce • compressing • consists • extends • struts

The suspension system

A basic suspension system **1** of springs and shock absorbers. There are several different types of springs such as coil springs, leaf springs, torsion bar springs and air springs. Springs are excellent at **2** energy when a car drives over a bump but not so good at dispersing it. Think of **3** a spring and then letting it go and you'll understand what that means. The force is almost uncontrollable and is why you also need absorbers.

Shock absorbers absorb some of the **4** so that the vehicle does not continue to go up and down just as a ball would until all the energy is gone and it just rolls. They are installed between the frame of the car and the **5**. When a car goes over a bump the spring first compresses and then **6** again. The force of the extension phase is very strong, so the absorber reduces the rebound considerably and controls the following up and down or side to side movements. Shock absorbers are often mounted inside springs – these are called **7**.

P **3** **Search the Internet for pictures of a strut, coil spring, leaf spring, torsion bar spring and air spring. Then prepare a presentation in which you use the pictures to explain the function of these different parts.**

→ **PHRASES:** Presentations

C | Drivetrain

The drivetrain is what transfers the power to the road surface to move the vehicle. It is also called the powertrain and is made up of all the parts of the vehicle that are necessary to produce the energy to drive the car, convert it into torque and then transfer it via the driveshaft, differential and drive axle to the wheels in order to turn them and so move the weight of the car.

1 Match the parts to their function. What are their German translations?

1. transmission 2. engine 3. differential 4. driveshaft 5. drive axle 6. clutch	a. produces the energy needed to drive a vehicle b. uses gears to reduce the high RPM of the engine c. sends the torque to the wheels via differential d. is depressed to disconnect the engine from the transmission e. enables the wheels to turn at different speeds f. the axle that is powered by the engine to drive the vehicle

2 Use five of the words in exercise 1 to label the picture.

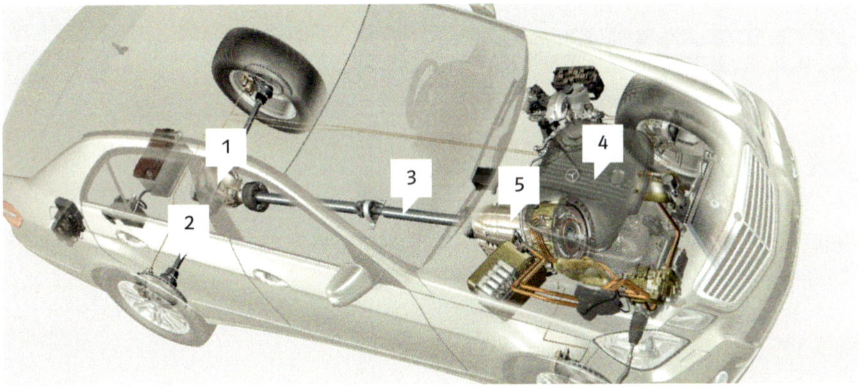

M **3** KMK II Ihr Vorgesetzter bittet Sie die Vor- und Nachteile von Fahrzeugen mit manueller oder automatischer Gangschaltung auf Grundlage der folgenden Auflistung in einem Fließtext auf Deutsch zusammenzufassen.

Automatic transmission	Manual transmission
• No clutch – torque converter instead • Fewer repairs • Easier to learn to drive – no stalling • Easier when driving in heavy traffic • Keep foot on brake when stopping, otherwise car will continue • Driver puts vehicle in drive and gear changes occur automatically • You only need one foot to drive • Must put vehicle in 'park' to leave car	• Cheaper to buy and to repair • Driver uses a gear stick to change gear and changes gears according to RPM • Driver feels in charge • Can consume less fuel if driven fuel-efficiently • Sportier and more fun to drive • You need both feet to drive • Can leave car in any gear

WORD BANK

air spring – Luftfeder
clutch – Kupplung
coil spring – Spiralfeder, Sprungfeder
compress (v) – zusammendrücken
driveshaft – Antriebswelle
extend (v) – (sich) ausdehnen
grip (v) – greifen, haften
leaf spring – Blattfeder
rebound – Rückfederung, Rückprall
RPM (revs / revolutions per minute) – Umdrehungen pro Minute
spring – Feder
strut – Federbein
torque – Drehmoment
torsion bar spring – Drehstabfeder

E

NACH DIESEM MODUL:

- √ kann ich eine Stellenanzeige auf Englisch verstehen,
- √ kann ich meine Fähigkeiten und Qualifikationen auf Englisch beschreiben,
- √ kann ich eine Bewerbung auf Englisch schreiben.

1

2

3

4

Applications

The world today is full of opportunities. You might decide to work in your hometown, a big city, or even another country. Maybe you want to work for a small family-run company or a major corporation with branches across the globe. Whatever you decide, the opportunities are out there, but it is up to you to take them.

1 **Look at the pictures above and describe them using the words below.**

> to apply • job advertisement • letter of application • CV • to search • job interview • employment contract • to invite • newspaper • to write

2 **Put the steps of the application process shown in the pictures above into the correct order.**

P **3** **Work on your own. What skills and qualifications do you need for a job in your industry? Make a list. Use these words to help you:**

> hard-working • motivated • logical • good with numbers • good with your hands • communicative • good team player • technical skills • able to understand plans and diagrams • attention to detail • computer skills

4 **Work with a partner and decide upon a final list. Share it with the class.**

A | Job advertisements

M/R **1** **Read the three online job advertisements below and find the English translations for the following German words and expressions.**

1. verantwortungsbewusst
2. aktiv, praktisch
3. Bewerber
4. Störungssuche
5. pünktlich
6. lernbereit
7. begeistert
8. gültige Fahrerlaubnis
9. Erfahrung
10. abgeschlossene Ausbildung
11. gute Bezahlung
12. fließendes Englisch
13. teamfähig
14. Montage
15. Grundkenntnisse in CAD
16. freie Stelle

ww#.totaljobs.com

1	City Garage – Car mechanic (full-time)	City Garage in Wolverhampton is looking for a recently-qualified car mechanic to join our busy team. Are you enthusiastic, willing to learn, responsible and punctual? If so, then contact us. We offer good rates of pay and a pleasant working atmosphere. Applicants must have a clean and valid driver's license. Please send a written application and a CV to Michael McCarthy at the following address: City Garage 17 Gorsebrook Rd. Wolverhampton WV6 5PI Tel.: 0044 (0)1902 554997...
2	Harper Shipping Client Support Technician	We are looking for a client support technician for our IT team to provide support and troubleshooting for hardware and software installation. You will also be responsible for new system configuration and maintaining the desktop client software. You should have two years experience, good English skills and be able to work as part of a team. Salary: negotiable Email application to: careers@HS...com
3	Trafford Tools Mechatronic Technician	Trafford Tools is a leading supplier of high-quality tools. We have a vacancy for a hands-on person to work as a mechatronic technician. Tasks: fault finding, maintenance, assembly, assist with installation. Skills: completed apprenticeship, basic CAD skills, good knowledge of tools, fluent English. Contact: James Truman J.truman@TT...co.uk Tel: 0044 (0)161 929496...

I/M **2** **One of your colleagues does not speak English and asks you to tell him about one of the job ads. Choose one of the ads and give him a brief summary in German. Swap roles.**

WORD BANK

branch – Filiale
family-run company – Familien-unternehmen
job application – Bewerbung
maintain (v) – warten
opportunities – Möglichkeiten
qualifications – Qualifikationen
skills – Fähigkeiten
troubleshooting – Fehlerbehebung

B | Letter of application

R **1** **Read the letter of application and decide whether the following statements are true or false. Give reasons for your decisions.**

1. Erkan is in the final year of his apprenticeship as a car mechanic.
2. He has always wanted to be a car mechanic.
3. Erkan had a holiday job at *Werkstatt Kuhn*.
4. He is looking forward to working abroad.
5. Erkan has included his CV in the application.

Erkan Celik • Waldstrasse 48 • 50679 Köln • Germany

Michael McCarthy
City Garage
17 Gorsebrook Rd.
Wolverhampton
WV6 5PI
England

12 June 20_

Dear Mr McCarthy,

Application for job as car mechanic

I am writing to apply for the advertised position of a car mechanic at City Garage. I completed my apprenticeship as a car mechanic at a technical college here in Germany six months ago. I have been interested in cars since I was a child and being a car mechanic is my dream job.
As you will see from my CV, I recently completed a six-month internship at Werskstatt Kuhn. This not only gave me the opportunity to put my theoretical skills into practice, but I also had the chance to gain more experience with customers.
I am a very enthusiastic person. I am also reliable, responsible and willing to learn. Furthermore, I have good English skills and I would welcome the opportunity to gain some experience working in another country.
Thank you in advance for considering my application. If you require any further information, please contact me.

Yours sincerely,
Erkan Celik

Encl.: CV

WATCH OUT!
I am interested ~~for~~ cars. I am interested <u>in</u> cars.

M **2** KMK I **Fassen Sie mit Hilfe der unten stehenden Stichworte die wichtigsten Inhalte des Bewerbungsschreibens auf Deutsch zusammen.**

1. Beruf, für den sich Erkan bewirbt **2.** Berufserfahrung **3.** Motivation
4. Persönliche Qualifikationen/Fähigkeiten **5.** Kontaktperson

C | The CV

R **1**
KMK II
A 1.4

Andrea Wagner ist Auszubildende bei einer IT-Firma in Nürnberg. Sie möchte sich für eine Praktikumsstelle in der englischen Niederlassung ihrer Firma bewerben und fragt David, einen Praktikanten aus England, ob er ihr ein paar Tipps geben kann. Hören Sie das Gespräch an und vervollständigen Sie die folgenden Sätze auf Deutsch.

1. Andrea möchte sich für eine Praktikumsstelle in ? bewerben.
2. Sie bringt am Montag ? mit ins Büro.
3. Der Lebenslauf sollte nicht länger sein als ? .
4. Man sollte niemals einen Lebenslauf ohne ? senden.
5. Man sollte den Lebenslauf dazu benutzen, seine ? und ? hervorzuheben.
6. Wenn man seine Arbeitserfahrung auflistet, sollte man mit ? anfangen.
7. In England muss ein Lebenslauf kein ? enthalten.
8 Man sollte sicherstellen, dass man zwei ? nennt.
9. Man sollte niemals ? .
10. Wenn der Lebenslauf fertig ist, sollte man ihn auf ? und ? überprüfen.
11. Es ist eine gute Idee, den Lebenslauf ? zu geben.
12. Wenn man den Lebenslauf abschickt, sollte man ? behalten.

P **2**
KMK II

Verfassen Sie ein Anschreiben für eine der in Modulteil A ausgeschriebenen Stellen auf Englisch. Arbeiten Sie dabei die unten aufgeführten Inhalte ein.

→ **PHRASES:** Applications

- Wählen Sie einen geeigneten Betreff.
- Nehmen Sie Bezug auf die Stellenanzeige und bekunden Sie Ihr Interesse an der ausgeschriebenen Stelle.
- Sie werden bald Ihre Ausbildung als … abschließen.
- Bringen Sie zum Ausdruck, dass Sie ein geeigneter Bewerber sind: organisiert, erfahren, hoch motiviert, teamfähig, flexibel, lernbereit, zuverlässig. Sie haben gute Englischkenntnisse.
- Beenden Sie den Brief angemessen und unterschreiben Sie.
- Weisen Sie auf Ihren Lebenslauf hin.

WORD BANK

consider (v) – betrachten
gain (v) – sammeln
internship – Praktikum
letter of application – Bewerbungsschreiben
reliable – zuverlässig
require (v) – benötigen

INFO BOX: Translating German school types and qualifications

It is not always possible to find the equivalent of your school type or qualification in English. However, the following table gives some useful translations.

Schools			
Grundschule	Elementary school / Primary school	Berufsschule (Duales System)	Part-time vocational school (dual system)
Hauptschule	General secondary school	Berufsfachschule	Full-time vocational school
Realschule	Intermediate secondary school	Fachschule für technische Berufe	Technical college
Gymnasium	Upper secondary school	Fachoberschule	Upper secondary vocational school
Gesamtschule	Comprehensive school	Fachgymnasium	Specialized upper secondary school

Qualifications			
Hauptschulabschluss	secondary modern school qualification	Fachabitur	advanced vocational certificate of education
Mittlere Reife	secondary school leaving certificate	Zusatzqualifikation	additional qualification
Abitur	A level/high school diploma		

06

NACH DIESEM MODUL:

- √ weiß ich, wie man die Prozessabläufe in der Kfz-Montage auf Englisch erklären kann,
- √ kann ich mich zum Bereich Logistik in der Kfz-Produktion auf Englisch äußern.

Assembly and logistics

After the car body has been painted, it moves to the final stage of the production process. It is taken to the assembly shop, which is where all the other parts are added to the body in a pre-defined sequence to produce the finished car. An efficient assembly shop process is heavily dependent on good logistics. Logistics have to make sure that every part is available in the right place, in the right quantity at precisely the time the mechanic needs to install or attach it to the vehicle. This process, called 'just-in-sequence', ensures a smooth and efficient assembly line.

1 Look at the picture of a car assembly line. Brainstorm together what you think is happening here.

2 Discuss why the doors of the cars have been removed.

3 Logistics are an important factor in car production. Write down three reasons and explain why. Then compare your results with another group.

4 Where does the car go when it leaves the assembly shop?

A | The assembly process

R **1** Complete the sentences with the correct form of the following words.

> adjust • assemble • check • fit • insert • install (2x) • insulate • lay • mount • put in • remove

1. The doors are … from the car body.
2. The car is … to the conveyor system.
3. The carpets and cables are … and the seats are … .
4. The cockpit is … on a separate line and then … on the main assembly line.
5. The electrical system is … to make sure everything works properly.
6. The doors are … carefully to make sure they close properly.
7. The door frames are … with rubber.
8. Most of the parts are … manually.
9. The lights are … to make sure the beam shines in the right direction.
10. The steering wheel is … into the steering column.

R **2** Listen to Mr Beck talking to a trainee about the first part of the assembly process.
⦾ A 4.6 Are the following statements true or false? Correct the false statements.

1. All the parts are added to the car frame in the assembly shop.
2. The car is not attached to a conveyor system.
3. All vehicle manufacturers use the same assembly process.
4. The chassis number only stays on the car as long as it is being produced.
5. After the doors have been removed from the car, they are taken to a separate door line.
6. All the parts are assembled on one of the main production lines.

R **3** Listen to Mr Beck talking about the rest of the assembly process and complete the
⦾ A 4.7 flow chart by matching the numbers in the chart with the words in the box below.

> adjust lights • put in seats • attach sunroof • install wiring harness • fit windshield and rear window • install doors • marriage • put in carpets • ready to ship • put in cockpit

stamp chassis no. ▸	remove doors ▸	1 ▸	2 ▸	electrical parts ▸	add battery ▾
put in edge protection ▾	◂ put in side panel	◂ 5	◂ 4	◂ fit headliner	◂ 3
install piping ▸	assemble tank ▸	6 ▸	add front module ▸	7 ▸	attach wheels ▾
drive check ▾	◂ align wheels	◂ 9	◂ door fit check	◂ pour in liquids	◂ 8
engine + paint check ▸	rain check ▸	10			

WORD BANK

assembly – Montage, Herstellung
body – Karosserie
conveyor system – Förderanlage
mount (v) – montieren, einbauen
steering column – Lenksäule

WORD BANK ⦾

align (v) – fluchten, ausrichten
edge protection – Kantenschutz
headliner – Dachverkleidung
hoist (v) – hochheben
fuel gauge unit – Kraftstoffanzeigeeinheit
manually – von Hand, manuell
piping – Verrohrung
power steering – Servolenkung
powertrain – Antrieb(sstrang)
pre-assembly – Vormontage
raise (v) – anheben
rolls test – Abrollprüfung
rubber sealant – Kautschukdichtungsmasse
seal (v) – abdichten, versiegeln
side panel – Seitenwand
windscreen wiper tank – Scheibenwischertank
wiring harness – Kabelbaum

B | The logistics department

A successful assembly shop relies heavily on logistics ensuring that everything is in the right place when the mechanic needs to install it. The logistics department also has to plan exactly what happens to any parts or raw materials in the plant; from the time they are unloaded on the docks and taken to the production line to the minute the final product leaves the plant. The material flow not only affects the workload of the employees but also the speed of production and the quality of the end product.

1 Match the logistics terms to the German translations.

1. aisle	a. Lieferant
2. dock	b. Rampe
3. dolly	c. Fahrstraße, Gang
4. stock	d. Lagerhalle
5. supplier	e. Lagerware
6. warehouse	f. Rollbrett

P **2** Make sentences in English to explain each of the terms in exercise 1.

3 Find possible word combinations. Then match three of the combinations to the pictures below and make a sentence about each picture.

1. bottle	a. container
2. collapsible	b. neck
3. conveyor	c. storage
4. forklift	d. bins
5. high-bay	e. crate
6. storage	f. belt
7. wire mesh	g. truck

1

2

3

P **4** Check the Internet and search for two more pieces of equipment used in logistics. Prepare an information sheet.

5 **UNDERSTANDING TECHNICAL ENGLISH**

Make sentences using these words in manufacturing and in an everyday context.

dolly • neck • aisle • stock • dock • belt • lift

C | Problems and solutions in logistics

R **1** ⓞ A 4.8 **Listen to the manager of a logistics department talking about some of the problems they had. Then match the problems to the solutions.**

1. The forklift drivers at the docks had either too much or no work to do.
2. Some parts were stored in eight different places.
3. There was not enough space for the goods.
4. Delays due to congested aisles.
5. Too many accidents in the aisles.

a. Loading routes were reorganised.
b. Maximum/minimum signs to control stock were put up.
c. Inventory was reduced.
d. The aisles were widened.
e. Trucks arrive continuously.

R **2** ⓞ A 4.8 **Listen again and write down what caused the problems.**

R **3** KMK II **Ihre Ausbilderin gibt Ihnen den folgenden Bericht aus einer englischen Fachzeitschrift und bittet Sie, einige Fragen dazu auf Deutsch zu beantworten.**

Improved logistics through global production

Originally German cars were produced in Germany, Japanese cars in Japan and so on. Today with an increasingly global market car companies have production plants all over the world. One of the advantages is that companies can easily compare their production facilities and see where they are producing most efficiently. This has led to competition between the plants with every plant manager wanting his or her plant to produce the highest quality vehicles at the lowest price.
Continuous improvement is the order of the day and as soon as a plant is not performing as well as others are, studies are carried out to determine where improvements can be made. Some of these studies have shown that keeping work processes simple is the key. Apparently the more decisions the line workers have to make, the more likely they are to make a mistake or waste time.

Once these findings became known, so-called 'kits' were introduced in some car plants. Kits are small sets of car parts that are pre-selected by workers before they go to the line. These parts are unpacked and put on narrow racks that are then taken to the line. That way each technician on the assembly line saves time and no longer has to decide which part to choose. And they do not have to walk to the large parts bins that contained enough parts for half an hour's production, take out the parts and unpack them. For the very small parts such as grommets, washers and screws, etc. they also introduced moveable racks (a kind of trolley that can be attached to the car as it moves along the conveyor). This again means less walking for the person on the assembly line.
Introducing kits was a success. It led to improved safety standards in the plant, a rise in vehicle production per person and considerable waste reduction.

1. Welchen Vorteil haben weltweite Produktionsstätten in der Kfz-Herstellung?
2. Mit welcher Zielsetzung werden Studien durchgeführt?
3. Warum sind komplexe Arbeitsgänge weniger effizient als einfache?
4. Aus welchem Grund wurden Montagesätze in Kfz-Werken eingeführt?
5. Weshalb sind diese Werke nach der Einführung von Montagesätzen sicherer geworden ist?
6. Was hat dazu geführt, dass die Produktivität gesteigert werden konnte?

WORD BANK

forklift – Gabelstapler
grommet – Halterungsöse
rack – Regal, Ständer
washer – Dichtung(sring)

WORD BANK ⓞ

delay – Verspätung, Verzögerung
inventory – (Lager-)Bestand

F

NACH DIESEM MODUL:

- √ kann ich *Small Talk* auf Englisch machen,
- √ kenne ich die Regeln des *Small Talk*,
- √ kann ich auf Englisch Essen erklären und bestellen.

1

2

3

4

Socialising

Socialising is very important. It can strengthen relationships with your colleagues and improve the working atmosphere in a company. It can also be a good way of getting to know your customers or potential customers a little better. And in today's global world, English has become the world's language. So although the people you meet and work with are likely to come from a wide range of cultural backgrounds, you will often be required to communicate in English. That means that it is important to not only have the necessary language skills, but also the intercultural knowledge so that you can make a good first impression and develop professional and personal relationships.

I **1** Look at the pictures above. What relationship do you think these people might have to each other? Discuss with your partner.

2 Brainstorm a list of topics these people might be talking about.

P **3** Choose one of the pictures and create a small dialogue.

4 Look at photo 1. Brainstorm what cultural misunderstandings could take place when a German employee meets a person from another country.

P **5** Make a list of situations when a German technician might have contact with foreign colleagues – for example, customers, suppliers, manufacturers, etc.

A | Small talk

Your company has visitors from a subsidiary in Poland. Fabian, a German trainee, has been asked to show Jan, a Polish trainee, around the company.

R **1** Listen to the dialogue between Jan and Fabian. Take notes of the topics the two are talking about. Then listen again and make a list of the phrases they use.
A 1.5

2 Name other topics that Jan and Fabian could also talk about.

3 Which of the following topics would / wouldn't you talk about? Why? Discuss in class.

free-time activities • films you have recently seen • sex • your job or school course • politics • money • personal problems • fashion • sports • food • the boss's wife • the weather • your favourite music • religion

P **4** Small talk is not that difficult. All that is needed is an 'ice breaker' to get things going. Look at the following examples and come up with appropriate answers.

→ **PHRASES:** Socialising

In the workshop:
a. Did you see the game last night?
b. How was your weekend?
c. It's warm today, isn't it?

In the car:
a. How was your trip / flight?
b. Is this your first time in Germany?
c. How's the weather in Sao Paolo?

At a trade fair:
a. Did you have any problems finding our stand?
b. Would you like a brochure?
c. If you like, I could give you a demonstration.

In the canteen:
a. Have you tried the tiramisu? It's amazing!
b. Is anybody sitting here?
c. Ah! I see you like spicy food.

P **5** Now it's your turn to break the ice! Think of at least three ice breakers for each of the following situations:

→ **PHRASES:** Socialising

1. You are at a bar with colleagues on Friday evening.
2. You work with a new colleague one afternoon.
3. You are at the "Green Technology" conference. It's time for a coffee break.
4. You've just arrived at work and you're in the lift. A person you don't know enters.

I **6** Choose one of the four situations above. Start a conversation and try to keep it going for as long as possible. Then end the conversation politely and act it out in front of the class.

→ **PHRASES:** Socialising

P **7** Check the Internet and create your own 'Dos & Don'ts' guide to making small talk and present it to your class. Also include some intercultural aspects.

Dos	Don'ts
Make eye contact.	Ask personal questions.

WORD BANK

subsidiary – Tochtergesellschaft
trainee – Auszubildende(r)

B | Small talk in business

R **1** KMK II **Ihr Arbeitgeber hat im Internet folgenden Artikel gefunden. Da er nur wenig Englisch kann, bittet er Sie, die folgenden Fragen auf Deutsch zu beantworten.**

1. Was wird oft in abwertender Weise über Small Talk gesagt?
2. Was passiert, wenn man sich nicht am Small Talk beteiligt?
3. Was sind die positiven Auswirkungen von Small Talk?

Small talk in business

Small talk is often dismissed as a waste of time, or as the art of saying nothing with a lot of words. Some people say it has no place in the business world, since the subjects discussed are often 'unimportant' – i.e. the weather, sports, fashion or other trivial matters. However, if you don't take part in small talk, your business partners might feel that you are being unfriendly or antisocial. When people just stand next to one another, without any signs of communication between them, it can lead to a feeling of uneasiness, and even tension. And this is certainly not good for future business or private relations. Small talk is in fact a very important and effective tool for avoiding an awkward silence between people who don't know each other, especially business partners who may want to form a relationship. Making small talk with strangers helps to break the ice; you discover what they like or dislike, so you can find a common ground of some sort. It helps, for example, if you support the same football team, like the same kind of car, or read the same kinds of books. However, it is also important to know who you are making small talk with. The rules of small talk are not international and different cultures have their own dos and don'ts. For example, you may have been told to always make eye contact when talking to your counterpart – but this may offend a material supplier in China. If you are going to have contact with someone from a different cultural background, it is a good idea to do a little research to avoid any embarassing mistakes.

M **2** KMK II **Ihr Arbeitgeber bittet Sie, für die Mitarbeiter stichwortartig ein kurzes Merkblatt anhand dieses Textes auf Deutsch zu erstellen.**

R **3** **Sind Sie ein guter Smalltalker? Bewerten Sie die Aussagen zum Small Talk in einem Online-Test mit „Yes" oder „No". Notieren Sie die entsprechenden Buchstaben.**

Statements	Yes	No
1. Small talk deals with rather unimportant topics. Another word for it is chitchat.	a	c
2. Religion always makes a good topic, because it is controversial.	d	e
3. Answering questions with 'yes' or 'no' during small talk is appropriate.	i	e
4. A good answer when asked if you'd like something to drink is: "I'd love a cup of coffee, thank you."	g	h
5. "That's great" is an acceptable answer to the statement: "I got the job!"	b	i
6. The noun "talk" can also mean speech.	a	c
7. Small talk is an icebreaker.	b	f

Addieren Sie die Buchstabenwerte a = 3, b = 3, c = 1, d = 0, e = 3, f = 1, g = 3, h = 1, i = 1.
Wer hat die höchste Zahl? (Maximum 21)

C | Eating out

1 Look at the German words (1.–16.) below and match them with the English equivalent (a.–p.).

1. Speisekarte	a. well-done
2. Rechnung	b. pork
3. Rind	c. side order
4. Vorspeise	d. bill
5. Schweinefleisch	e. game
6. Wild	f. dessert
7. durchgebraten	g. savoury
8. Hauptgericht	h. fork
9. Geflügel	i. menu
10. Kellner(in)	j. main dish
11. Nachtisch	k. beef
12. Trinkgeld	l. poultry
13. Beilage	m. starter
14. vegetarisch	n. waiter / waitress
15. herzhaft	o. vegetarian
16. Gabel	p. tip

I/M **2** Partner 1: You are taking a foreign colleague out to dinner in a traditional German restaurant. Study the menu below and explain the dishes to him / her.
Partner 2: Take the role of the foreign colleague and ask questions about the menu.

Gasthaus Sonne

Tagesmenu (10,95 €)

Vorspeisen

Lauchcremesuppe **oder** *kleiner gemischter Salat*

***

Hauptgerichte

Rinderrouladen mit Kartoffelklößen & Sauerkraut

oder

Putenschnitzel mit Pommes Frites

oder

Forelle Müllerin mit Salzkartoffeln

***

Nachtisch

Rote Grütze mit Sahne **oder** *Gemischtes Eis (Vanille / Erdbeere / Schoko)*

Inklusive ein alkoholfreies Getränk (0,3 l)

WORD BANK

avoid (v) – vermeiden
awkward – unangenehm
common ground – gemeinsame Basis
dismiss (v) sth. as – etwas abtun als
tension – Spannung
trivial – unbedeutend
uneasiness – Unbehagen

I **3** Now the two of you are ready to order. The young German waitress / waiter doesn't mind taking the order in English. Study the eating out phrases at the end of the book and create a role play. Practise, and perform it in front of the class.

→ **PHRASES:** Socialising

07

NACH DIESEM MODUL:

- ✓ kann ich auf Englisch wichtige Teile des Motors beschreiben,
- ✓ kann ich mich auf Englisch zu Verbrennungsmotoren, Dieselmotoren und Turboladern äußern.

Engine basics

Most cars on the roads today still run on petrol or diesel and are powered by the standard combustion engine, often referred to as the Otto engine after the inventor Nicholas August Otto. Although many improvements have been made to the first Otto engine, the basic principle is still the same.

1 Match each engine part to its function.

1. camshaft
2. connecting rod
3. crankshaft
4. flywheel
5. piston
6. spark plug
7. sump
8. valve
9. V-belt

a. metal part that moves up and down in the car cylinder
b. shaft with cams which control the opening and closing of the valves
c. changes the piston's linear motion to a rotational motion
d. joins the piston to the crankshaft and rotates at both ends
e. opens and closes to let air or liquid in or out
f. ignites the fuel air mixture
g. a round band that links two or more rotational shafts
h. a reservoir of oil for the lubrication
i. revolving device attached to the crankshaft that stores energy

2 Which parts in exercise 1 are shown in the pictures above?

3 Search the Internet for pictures of a combustion engine. Choose your favourite picture and label the different parts using the words from exercise 1.

A | How a basic combustion engine works

R **1** Complete the text with the words from the box and find out how combustion works.

valve • ignites • combustion • crankshaft • connecting rod • induction • compression • cylinder • wheels • piston • exhaust • fuel

THE COMBUSTION ENGINE

The internal combustion engine burns fuel and air to convert it into the energy that is then used to rotate the **1**. The main part of the engine is the **2**.
Car engines mostly have 4 cylinders although larger engines can have 6, 8 or even 12. Each cylinder contains a **3**. The piston moves up and down inside the cylinder. At the top of the cylinder there are two valves (or sometimes four) – one intake valve and one outlet, or **4** valve. The piston itself is cylindrical. It is usually made of steel or aluminium alloy and has rings on the outside. These rings form a sliding seal so that when the piston moves up or down, no **5** can escape into the sump and no oil can enter the **6** chamber. The piston is attached to the crankshaft via the **7**.
Most engines are four-stroke engines. During the first stroke, the **8** stroke, the piston goes down, the inlet **9** opens, a mix of petrol and air is either sucked or injected into the combustion chamber and the valve closes again. During the second stroke the piston is pushed up again. As the piston rises, it compresses the fuel air mix. This stage is known as the **10** stroke. Immediately after the piston has reached top dead centre, the spark plug produces a spark, which **11** the fuel air mix in the combustion chamber. That is known as the combustion stroke. The heat causes the air to expand pushing the piston back down. The energy from the combustion turns the axial movement to a rotational movement which travels via the **12** into the transmission and then to the wheels. Then, in the fourth stroke, the outlet or exhaust valve opens as the piston rises again pushing out the gases which remain from the burning fuel mixture, known as the exhaust. The four-stroke cycle is then completed and starts all over again.

2 Which picture shows which type of stroke: combustion stroke, exhaust stroke, compression stroke or intake stroke?

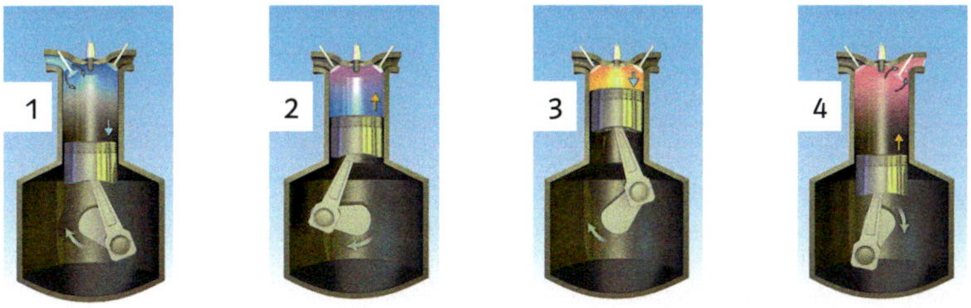

R **3** Listen and answer the questions below in complete sentences.
A 4.9

1. How are the cylinders of the boxer engine arranged?
2. What is the boxer motor also known as?
3. How many cylinders does the V8 engine have?
4. How are the cylinders of a V6 engine arranged?

WORD BANK

alloy – Legierung
cam – Nocken
camshaft – Nockenwelle
combustion chamber – Brennkammer, Verbrennungsraum
combustion engine – Verbrennungsmotor
compression stroke – Druckhub
connection rod (conrod) – Pleuelstange
crankshaft – Kurbelwelle
exhaust valve – Auslassventil
flywheel – Schwungrad
four-stroke cycle – Viertakt-Zyklus
ignite (v) – zünden
induction stroke – Ansaugtakt
intake valve – Einlassventil
piston – Kolben
shaft – Welle
sliding seal – Gleitdichtung
spark plug – Zündkerze
sump – Ölwanne

WORD BANK

horizontally opposed – Boxeranordnung; horizontal gegenüberliegend

B | The diesel engine

R **1** What do you know about diesel engines? Do the quiz and say which statements are true or false. Then listen and check your answers.
A 4.10

1. A diesel engine doesn't have a spark plug.
2. A diesel engine uses less fuel than a petrol engine.
3. Diesel fuel is not as refined as petrol.
4. The combustion chamber of a diesel engine is always pre-heated.
5. The compressed air in the combustion chamber of a diesel engine reaches a higher pressure than the fuel-air mix in the petrol engine.
6. The more air is compressed, the hotter it gets.
7. Fuel is injected into the compression chamber at the same time as air is drawn in via the inlet valve.
8. All diesel engines have a glow plug.
9. Diesel engines are cheaper than petrol engines.
10. Diesel evaporates more quickly than petrol.

QUIZ

10 correct answers:	Well done, you are an expert!
8 – 9 correct answers:	You are pretty good!
5 – 7 correct answers:	Listen again and do the quiz once more.
1 – 4 correct answers:	Maybe you are just having a bad day?

2 Match the English expressions to their translations.

1. to shift	a. Pferdestärke (PS, am Rad gemessen)
2. acceleration	b. turboaufgeladen
3. brake horsepower (bhp)	c. Leistung
4. displacement/cubic capacity	d. Nennleistung
5. manual transmission	e. Benzinmotor
6. performance	f. Drehmoment
7. petrol engine	g. Schaltgetriebe
8. rated output	h. schalten
9. torque	i. Beschleunigung
10. turbocharged	j. Hubraum

R **3** Use the words from exercise 2 to complete Frank's email below.

Hi Robert,
Guess what? I've bought a new car! This one has much better **1** than the last. It's a proper V6 engine, so 6 cylinders, with a **2** of about 168 bhp! What's more, I've switched to a **3** and the **4** is great! Nought to sixty in 8.2 seconds! Of course, with approximately 240 Nm it has less **5** than the diesel had. But the engine size more than makes up for it. The **6** is about 3100 cc, which is really quite big. However, the V6 engine only comes as an automatic and I do miss **7** with the gear stick. Driving a car with a **8** transmission always made me feel like a bit of a racing driver. But then my old diesel engine wasn't **9** so I can really tell the difference on the motorway. The **10** at high speeds is much better in my new car.
If you're free this evening, why don't you come round? You'll be surprised when you see the brand!
Cheers, Frank

C | Turbo chargers

R **1** Read the text and list the advantages and disadvantages of turbo chargers.

ww#.turbochargers_info.uk

FAQs

What exactly is a turbo charger and how does it work?

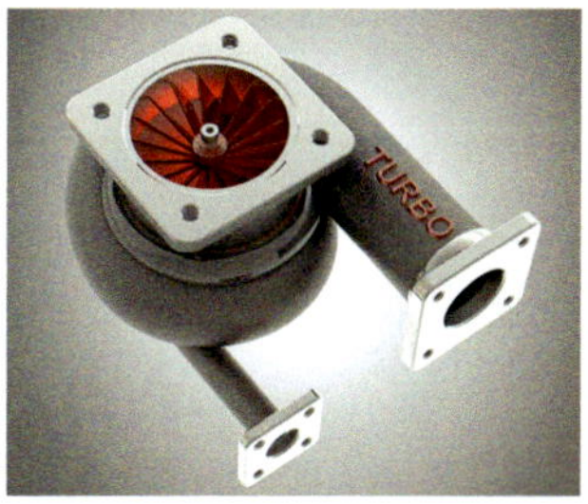

Alfred Büchli received the first patent for a turbo charger in 1905 when they were used in ships. Today turbo chargers are the basis for downsizing – an expression that means to get more power out of smaller engines in order to save fuel and reduce the CO_2 level. That way engines can also be lighter because you are able to produce the same power with less cubic capacity.

A turbo charger uses the waste energy from the exhaust gas to generate more power from the internal combustion engine. The exhaust gas drives a turbine in the turbine housing. The turbine wheel is connected to a compressor wheel by a shaft. As the compressor wheel spins, fresh air is drawn in, compressed and then directed to the combustion chamber. Compressing air gives you more air molecules in the combustion chamber, more fuel can be injected so you get more combustion energy.

One problem is the very high temperatures in the exhaust gas stream. In some engines this can reach over 1100°C, which will melt most metals. The other is dealing with what we call 'turbo-lag'. That is the delay caused by the time it takes to bring the turbo up to speed. At the moment some cars are fitted with two turbo chargers, one smaller one and one larger one to try to prevent this delay. It is a lot better now.

P **2** Search the Internet to find out what 'common rail technology' is and how it works. Prepare a short presentation to give to the class.

→ **PHRASES:** Presentations

M **3** KMK II Ihr Teamleiter bittet Sie, folgenden Artikel für Ihre Kollegen tabellarisch darzustellen. Lesen Sie den Artikel und fügen Sie die Informationen in eine Tabelle auf Deutsch ein. Schreiben Sie danach eine kurze Zusammenfassung auf Deutsch.

NEW SERIES OF TURBO-DIESEL ENGINES

Recently, one of Europe's largest car manufacturer's launched a new series of turbo-diesel engines. It is the second generation of a multi-jet, multi-valve turbodiesel power unit, with a displacement reduced to 1600 cc, down from the 1900 cc of the first generation. Based on the same design, the new 16-valve engine has an 80 mm cylinder bore and a stroke of 81 mm. The old generation was equipped with a fixed geometry turbo charger and developed a peak torque of 290 Newton meter (Nm) at 1500 revolutions per minute (rpm), and 103 brake horsepower (bhp) or 77 kW at 4000 rpm. The more powerful second generation is now equipped with a variable geometry turbo, and delivers a torque of 300 Nm at 1500 rpm, and 88 bhp at 4000 rpm. Compared to the former 1.9 multi-jet, 8-valve power unit, the new engine is a considerable step forward. The new design of the combustion system and the increased pressure of the multi-injection system at 1600 bar, instead of 1300 bar, have both considerably improved overall performance, comfort and emissions. In addition, the architecture of the new engine generation includes an integrated diesel particulate filter and a catalytic converter, which are directly connected to the exhaust system. The new 1.6 turbo-diesel generation is going to be built into all their new models.

WORD BANK

catalytic converter – Katalysator
cylinder bore – Zylinderbohrung
displacement / cubic capacity – Hubraum
gear stick – Schalthebel
glow plug – Glühkerze
particulate filter – Partikelfilter
spin (v) – (sich) drehen
turbo charger – Turbolader

WORD BANK ◎

mileage – Reichweite, Benzinverbrauch

G

NACH DIESEM MODUL:

- ✓ kann ich eine englische Präsentation vorbereiten,
- ✓ kann ich auf Englisch präsentieren.

1

2

3

4

Presentations

At some point in your career you may be required to present your company, or their products or services, either informally or on a more formal level. This could mean presenting a new product to a potential customer, or presenting the company as a whole, or something more specific, such as assembly instructions or a positional sketch. Whatever you may be presenting, the ability to give an oral presentation is a key skill. It is also a skill that can be developed and improved.

P **1** **Look at the pictures above and describe them using the words below.**

to present • meeting • flipchart • graph • chart • laptop • preparation • to explain • sales pitch • audience • projector

P **2** **For what reasons might you have to give a presentation? Write a list.**

3 **With a partner, decide upon a final list and share it with the class.**

4 **With the same partner brainstorm what you think is important to remember when giving a presentation. Write your answers in a word web like the one below.**

A | Preparing a presentation

1 **In the following table are words which are linked to presentations. Match the English words (1.–10.) with the appropriate German words (a.–j.).**

1. audience	a. Augenkontakt
2. prompt cards	b. beschreiben
3. body language	c. Zusammenfassung
4. eye contact	d. visuelle Hilfsmittel
5. handout	e. Publikum
6. to prepare	f. Schaubild
7. graph	g. Stichwortkarten
8. visual aids	h. Körpersprache
9. to describe	i. vorbereiten
10. summary	j. Informationsblatt

M **2** KMK II **Fassen Sie den folgenden Text über die Vorbereitung einer Präsentation unter Beantwortung folgender Fragen auf Deutsch zusammen.**

1. Warum soll eine Präsentation nicht schriftlich ausformuliert werden?
2. Wie soll eine Präsentation gegliedert sein?
3. Was ist „signposting“?
4. Welche Vorteile bieten visuelle Hilfsmittel?
5. Wie müssen Präsentationsfolien aussehen, damit sie wirken?

How to make a presentation

You have been asked by your boss to give a presentation? Here are a few tips:

→ Do not write out your presentation in full. Instead, use numbered prompt cards with key words – they are the best way to remember important points.
→ Remember: your presentation needs to have an introduction, a main part, and a conclusion.
→ Begin by giving a brief overview of the points to be covered.
→ Use "signposting" phrases to lead through your presentation, e.g.:

> I would like to begin by ... • Let me start by showing you ... • Next, I would like to say something about ... • Another important point is ... • Finally, I will talk about ... • In conclusion, ...

→ Always finish with a conclusion.

Visual aids

You can use visual aids, such as Powerpoint, to help reinforce what you say. Visual aids help you explain complicated ideas or technical details more easily and hold the interest of your audience. They also make your presentation look more professional.
If necessary, you can print them out and use them as a handout for the audience.
Visual aids could include graphs, pictures, plans, diagrams, or short texts.
However, if you do use your visual aids to show text, then follow these guidelines:

→ Limit the text to six lines.
→ Try to use no more than six words per line.
→ Print the text in large letters, using upper and lower case letters.
→ Use dark colours, such as black, red, blue, or green. Light colours are difficult to read.

WORD BANK

assembly – Montage
positional sketch – Technologieschema

3 Match the visual aids (1. – 6) with the pictures below (a. – e.).

1. plan 2. pie chart 3. line graph 4. bar chart 5. diagram 6. table

a.

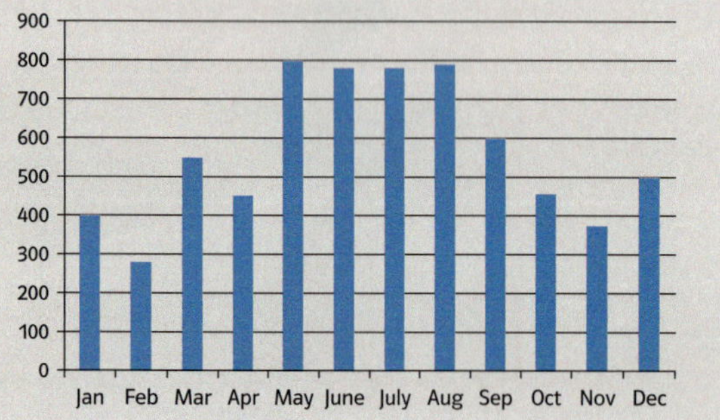

b.

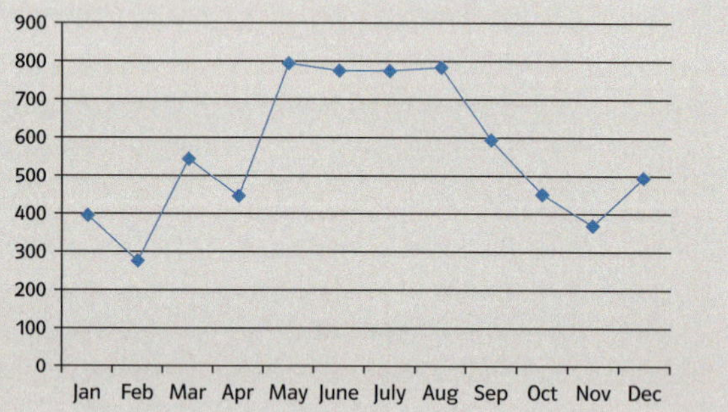

c.

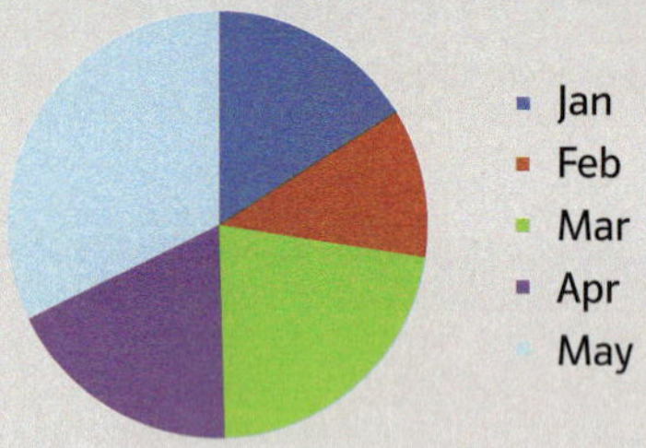

d.

2012		
Jan	$1,256	-6%
Feb	$1,477	+4%
Mar	$1,288	-5%
Apr	$1,543	+9%

e.

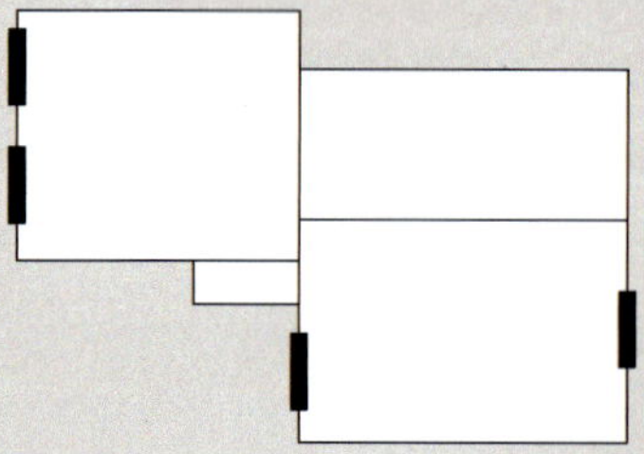

f.

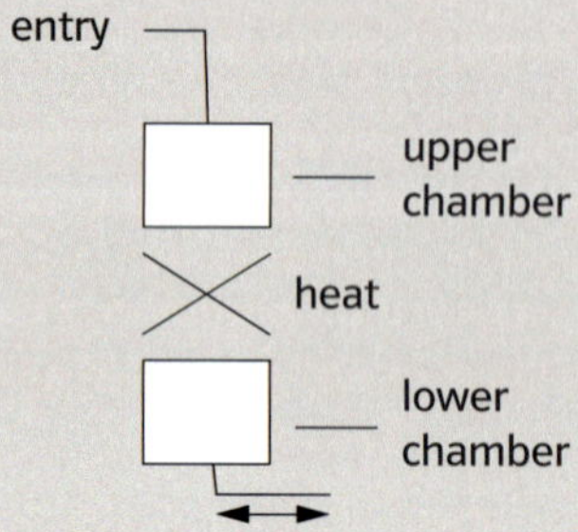

B | Describing materials and products

P **1** Think of something that you often use and describe it by using the terms in the box to help you. Don't name the item you are describing. Try and write 3 to 5 sentences.

I **2** Take it in turns to read your sentences to each other and see if you can guess what your partner is describing.

INFO BOX: Useful phrases for describing materials and products

It is a	strong · sturdy · rigid · flexible · lightweight · shiny · robust · matt	material.
The design is	modern · state-of-the-art · classic · elegant · old-fashioned.	
It is	user-friendly · difficult to operate · reliable · innovative · hand-crafted.	

C | Delivering a presentation

R **1** ◎ A 1.6 **Max Wiesinger, a trained carpenter who works for a furniture design company, is presenting a new line of furniture at a trade show. Listen to the presentation and complete the following text.**

Good morning ladies and gentleman. Today **1** our new range of **2** , *Simply Wood*. The range, which is shown here on the first slide, includes chairs, a large table, a lounger, and benches. I would like to start by talking about the **3** used in the furniture. All of the *Simply Wood* range is made from single softwood timber and has been **4** by our highly-skilled carpenters and joiners.
If you look at this picture, you can see that the benches and the seats are designed with a low back, deep seats and long arms to offer a great deal of comfort and convenience. Thanks to a special varnish, the furniture is 100 % **5** and will not suffer from discolouring. **6** from this final slide, the furniture is also available in white, to give it a classic, timeless look. And do not worry, we have used a durable white polyurethane coating that is suitable for every environment.
On your handout you will see more details about each piece, including the measurements. **7** this morning. Are there any questions? Yes …

P **2** **Now it's your turn. Work in small groups and design a chair, a pocket torch, or any other object of your choice. Present your ideas in class. Use the following words and phrases to help you:**

→ **PHRASES:** Presentations

We have designed a … • Our idea was to … • Our design is made of … • The advantage of the material we used is … • Its measurements are … • length • diameter • weight • width • height • rectangular • circle • circular • square • smooth edges • glued/screwed together • made out of one piece

M **3** KMK I **Ihre Kollegin muss nächste Woche eine wichtige Präsentation halten und bittet Sie um Ihre Hilfe. Sie hat den folgenden Text gefunden, versteht aber leider nur wenig Englisch. Übertragen Sie den Text für sie sinngemäß ins Deutsche.**

7 Ways to Do it Right: Presentations

1. Keep it short and simple (KISS).
2. Do not use too many visual aids.
3. If you are presenting in a group, decide beforehand who will say what.
4. Practice at least once in front of an audience.
5. Do not read from your notes. Look at your audience.
6. Structure your presentation with phrases so your audience can follow more easily.
7. Tell the audience when they can ask questions – either during the presentation, or at the end.

WORD BANK

carpenter – Tischler(in) / Schreiner(in)
circle – Kreis
circular – kreisförmig
diameter – Dicke
furniture – Möbel
height – Höhe
length – Länge
rectangular – rechteckig
smooth – glatt
square – quadratisch
trade show – Messe
weight – Gewicht
width – Breite

WORD BANK ◎

durable – haltbar
joiner – Tischler(in) / Schreiner(in)
polyurethane coating – Polyurethanbeschichtung
softwood – Weichholz
timber – Holz
varnish – Lack

08

NACH DIESEM MODUL:

- ✓ kann ich mich zu Brennstoffzellen, Hybridautos und elektrischen Autos auf Englisch äußern,
- ✓ weiß ich, wie ich einige Umweltthemen auf Englisch präsentieren kann.

1

2

3

4

Alternative engines

As we run out of fossil fuels such as gas and oil, car manufacturers are working hard to increase the efficiency of their combustion engines and to find other forms of energy. There are three main new engine technologies that are already on the road today while research and development continue: the fuel cell, which runs on hydrogen, the battery-operated electric car and the hybrid, which is a combination of the new battery-propelled and standard combustion engine.

1 Match the pictures to the correct information below.

a. This car has a fuel tank.
b. This car has two engines.
c. This car uses petrol.
d. This car uses hydrogen.
e. This car never has to be taken to a filling station.
f. This vehicle emits H_2O.
g. This vehicle emits CO_2.
h. This vehicle is only good for short distances.

2 **SAFETY FIRST** ⚠

Search the Internet and find out what harm poisonous emissions can cause to people and the environment. Make a list of the problems and their causes. Present the results to the class.

→ **PHRASES:** Presentations

A | The electric vehicle

1 Complete each sentence by putting the jumbled letters of the word in brackets in the correct order. Then write down the complete sentence.

1. The … are the gases that come out of the exhaust. (SOESIMINS)
2. An engine that burns petrol is called a … engine. (BCIMNOSTUO)
3. A car that runs on batteries is an … car. (CEECILTR)
4. The C in $C0_2$ stands for … . (ABCNOR)
5. The … is the place in the car that stores the fuel. (AKNT)
6. When we talk about the … we mean everything to do with our surroundings. (EEIMNNORTNV)

R **2** Read the article below to check your answers in exercise 1.

The return of the electric car

It is difficult to believe, but this is not the first time we have seen electric cars on our roads. At the end of the 19th century most of the cars were electric and the internal combustion engine was seen to be the technology of the future. Today it is exactly the other way around. Nearly all of the vehicles on the roads nowadays are propelled by an internal combustion engine and we look to the car with an electric engine that gets its power from rechargeable batteries via a controller as a major solution for the future.

So what has led to the revival of the electric vehicle, or EV as it is now often called? Well, first, we are running out of fossil fuel and secondly, combustion engines burn petrol, which makes them dangerous to the environment and to people. That is because they emit substances such as carbon dioxide, carbon monoxide, nitrogen oxides, sulphur oxides, and hydrocarbons. The electric car on the other hand, has zero emissions, it has no fuel tank and it does not have an exhaust pipe. EVs are becoming more popular and some companies are even making kits to convert normal cars into electric cars.

R **3** According to the article, what reasons are there for building electric cars today?

P **4** Search the Internet and find out how to convert a traditional petrol-driven car into an electric car. Write down a set of instructions to present to the class.

5 UNDERSTANDING TECHNICAL ENGLISH
The words 'charge' and 'pipe' can be used in different ways. Write sentences and use each word with a technical and an everyday meaning.

WORD BANK

carbon dioxide – Kohlendioxid
combustion engine – Verbrennungsmotor
controller – Steuergerät, Regler
emit (v) – ausstoßen
exhaust – Auspuff
fuel cell – Brennstoffzelle
fuel tank – Treibstofftank
hydrocarbon – Kohlenwasserstoff
hydrogen – Wasserstoff
kit – Ausrüstung, Set
nitrogen – Stickstoff
propel (v) – antreiben
rechargable – wieder aufladbar
sulphur – Schwefel

B | The fuel cell car

1 Match the terms on the right with the numbers in the picture.

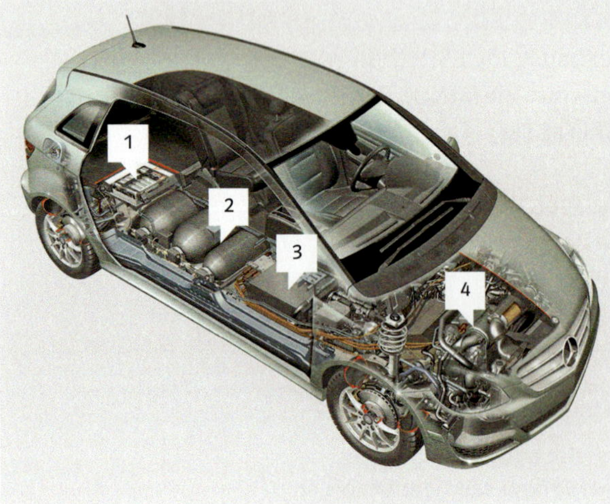

a. hydrogen tanks
b. electric motor
c. lithium-ion battery
d. fuel cell stack

R **2** Listen and make notes on the difference between an electric car and a fuel cell car.

A 4.11

3 Match each of the phrasal verbs to its synonym.

1. get back to sb.	a. go, leave
2. get down to sth.	b. contact again
3. get going	c. become familiar with
4. get on with sb.	d. pass through
5. get through	e. start
6. get used to	f. understand, like

R **4** Read Bob's email and decide which words fit best. Some words are used more than once and sometimes none is needed.

around • back • behind • down to • from • going • in • into • off • on • out • used to

Hi Frank
How are you? I promised to get **1** to you and let you know how things are going. Well, the course got **2** to a good start. I'm really getting **3** this new technology stuff. I've learned so much that my head is spinning.
I get **4** really well with one of the other apprentices, a guy called Peter. Today we managed to get **5** one of the trainers and he let us test drive the electric car. It was brilliant. We only had classes in the morning, so we got **6** at 12:30 and the car was waiting for us. I got **7** the wheel as Peter doesn't have a driving licence. As soon as we got **8** of town we drove straight to the motorway. When we got **9** there, we could really test the car. Driving an electric car actually takes a while to get **10** because the accelerator reacts immediately. The torque is amazing – we got **11** 0 to 100 in no time at all and it is so quiet! No wonder people get **12** so many speeding tickets driving those things.
I must get **13** now as I have to get **14** some work. See you when I get **15**.
Bob

C | The hybrid car

1 Complete the text about hybrids using the terms in the picture.

A hybrid vehicle contains parts of both petrol-driven and electric vehicles. It combines an electric engine with a conventional **1** propulsion system which gets its electricity from a rechargable **2**. This is recharged either with power from the combustion engine or by getting energy while slowing the car down – in which case the engine takes on the function of a generator. The **3** help increase the performance and efficiency of the vehicle. The **4** helps cool the engine.

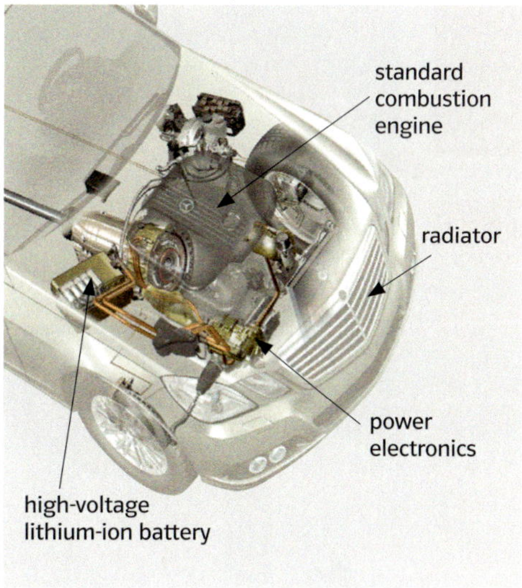

2 Read and complete the text by choosing the best verbs from the box below and putting them into the most suitable tense.

see • accelerate • drive • hear • know • say • utilise • use

If you look up the word 'hybrid' in the dictionary, it **1** something like this: 'a hybrid is a mixture, it is something that is made by combining two different things'. And that is exactly what a hybrid car is: a kind of mix. Almost everyone **2** what a normal car with an internal combustion engine is and you might **3** about electric vehicles a lot recently. Perhaps you **4** an electric car or maybe you know someone who **5** one. A hybrid is a combination of these two cars. It is a car which is fitted with two engines: one small petrol engine and one electric engine.
In the meantime there are different types of hybrids but the traditional hybrid car **6** both engines to propel the wheels. The electric engine **7** the car and then switches off so that the petrol engine keeps the car running while it is cruising. During that time the electric motor uses no energy.
Other hybrids only **8** the petrol or diesel engine to charge the electric engine and the batteries so that the car never has to be plugged in. Fuel consumption can then be very low and some hybrids can drive 20 km on just one litre of diesel.

R **3**
KMK II
◎ A 4.12

Ihr Abteilungsleiter übergibt Ihnen eine CD auf Englisch und bittet Sie, die Vorteile und Nachteile von verschiedenen Antriebssystemen zu notieren und auf Deutsch in folgende Tabelle zu übertragen.

Verbrennungsmotor	Brennstoffzelle	Elektromotor	Hybrid

WORD BANK

accelerator – Gaspedal, Beschleuniger
fuel cell stack – Brennstoffzellen-stapel
propulsion system – Antriebssystem
torque – Drehmoment

WORD BANK ◎

air pollution –Luftverschmutzung
compressed – komprimiert
flammable – leicht entzündlich
pedestrian – Fußgänger(in)
refuel (v) – nachtanken
regenerative braking – Brems-energierückgewinnung
run out of sth. (v) – ausgehen

H

NACH DIESEM MODUL:

- ✓ kann ich auf Englisch auf Kundenbeschwerden reagieren,
- ✓ kann ich auf Englisch mit unzufriedenen Kunden umgehen.

1

2

3

4

Dealing with customers

The way a company deals with its customers is the key to its success. Whether you are arranging appointments, handling enquiries, or dealing with complaints, it is essential that you are able to express yourself clearly, confidently, and politely. And remember: the customer may not always be right, but they are very important.

1 **Brainstorm words that describe what is shown in the photos above.**

P **2** **Compare your list of words with your partner and work together to describe the pictures. Use these words to help you:**

complaint • hotline • builder • hard hat • building site • meeting • customer • plans • mechanic • explain • repairs • bonnet • overalls • engine

3 **Which of the customers above do you think are happy, and which are unhappy? Explain your answer.**

P **4** **In what situations do you deal with customers in your job? Copy and complete the word web below to brainstorm reasons for talking to customers.**

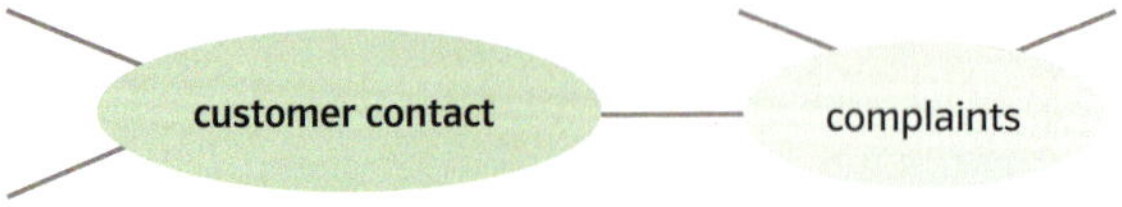

A | Complaints

1 Match the common causes for complaint on the left with the suggested solutions on the right. Several options may be appropriate.

1. delay in delivery	a. offer a price reduction
2. faulty goods	b. improve the service
3. damaged goods	c. repair the goods
4. unsatisfactory services	d. replace the goods
5. wrong goods	e. send the goods by air freight
6. too many/too few goods received	f. send the missing goods
7. too early delivery	g. take the goods back
	h. collect the surplus goods
	i. send a credit note

P **2** What about your industry? Which are the most common reasons for complaints? Make a list and compare it with a partner. Be as specific as possible.

R **3** Which of the following phrases are used for making complaints and which ones for replying to complaints? Make a list.

→ **PHRASES:** Dealing with customers

1. This is very inconvenient for us because we need the goods urgently.
2. We are sorry for any inconvenience this has caused.
3. We will keep the damaged goods until we hear from you.
4. Please return the damaged goods at our expense.
5. We can assure you that this will not happen again.
6. Some items were badly damaged.
7. We regret to inform you that the order has not yet arrived.
8. We think the damage may have occurred in transit.
9. We are experiencing difficulties with our new logistics software.
10. Could you please look into this problem immediately?
11. Please accept our apologies.
12. The reason for the mistake is that we are extremely busy at this time of the year.
13. Please send us replacements as soon as possible.
14. We hope that this proposal will find your approval.
15. We believe the problem is down to human error.

R **4**
KMK II
◎ A 1.7

You work for *Deutsches Netz*, an IT company in Hamburg. Your boss, Mr Bauder, receives a voicemail message on his mobile phone from Mr Jenkins, owner of *New Wave Marketing*, a subsidiary of an English marketing company based in Hamburg. Listen to the message and answer the following questions.

1. What did the technicians from Deutsches Netz do for New Wave Marketing?
2. What problems are the employees having (three things)?
3. What is the most serious problem?
4. Why is this a serious problem?
5. How should Mr Bauder contact Mr Jenkins?

WORD BANK

bonnet – Motorhaube
builder – Bauarbeiter
building site – Baustelle
hotline – Informationsdienst
repairs – Reparaturen
technician – Techniker(in)

P **5** **Mr Bauder asks you to respond to Mr Jenkins' phone call. He sends you the following email and asks you to draft a short email to the customer.**

→ **PHRASES:** Dealing with customers

Von: mailto:s.bauder@DNetz...de
An: mailto:info@DNetz...de
Betreff: Bitte an Herrn Jenkins schreiben (eilt!!)

... diese Firma ist ein sehr wichtiger Kunde von uns und wir wollen ihn **auf keinen Fall** verlieren!! Bitte sagen Sie ihm folgendes:

- Es tut uns sehr leid, dass es diese Probleme gegeben hat.
- Wir verstehen völlig, dass es jetzt sehr ungelegen und ärgerlich ist.
- Wir vermuten, dass das Problem durch menschliches Versagen verursacht wurde.
- Wir werden sofort jemanden schicken, um das Netzwerk zu reparieren. (selbstverständlich auf unsere Kosten!)
- Der Techniker wird spätestens um 17 Uhr dort sein.

Bringen Sie nochmals unser Bedauern für das Problem zum Ausdruck und versichern Sie ihm, dass sich dies nicht wiederholen wird.

Danke,
Bauder

M **6** **You work for a plumbing and heating installation company in Leipzig. Your supervisor, Mr Hüber, receives an email about a recent job done for an American family living in the area. Your supervisor does not speak very good English and asks you to read the email and to sum it up in German.**

Von: mailto:j.t.levine@kwq...net
An: mailto:info@sippelheizung_sanitaer...net
Betreff: Problems with the heating

Dear Mr Hüber,

I am writing in reference to the heating system installed by your company at my house last week. Unfortunately, my family and I have been having some problems. Firstly, the water in the bathroom takes a long time to warm up – sometimes over 5 minutes!

Secondly, the temperature in the house is too low, and I cannot make it warmer. Also, I cannot find the instruction manual – did your technician take it with him? My final problem is that the underfloor heating does not seem to be working in the bathroom. The tiles on the floor are still cold in the morning. Is something wrong?

As it is winter, and very cold, I would ask you to send someone to my house as soon as possible to find a solution to these problems. Please give me a call, or send me an email to confirm.

Thank you in advance,

John Levine

B | Customer service

P **1** What five tips would you give a colleague about dealing with unhappy customers?

2 Compare your list with a partner.

How to deal with unhappy customers

No company has happy customers 100 % of the time.
Things can always go wrong and customers can become unhappy – or worse, upset. Here are some tips on how to avoid making the situation worse.

Stay calm: The customer may be angry at somebody, but it does not help the situation if you get just as angry. The last thing you want to do is start an argument – that is very unprofessional and unhelpful. Try to calm the customer down.

Listen, listen, listen: You will only be able to understand what has upset the customer if you listen, and listen carefully. Before you start talking, listen actively to the customer until you fully understand why they are unhappy. If you don't know what's wrong, you will not be able to do anything about it.

The customer is king: Show the customer that they, and their problem, are important to you. Tell them quickly how sorry you are for their problem and reassure them that you will do everything you can to help them. And, always be patient and polite.

Stay positive: Do your best to stay positive at all times. Remember that your goal is to solve the problem and you can do this better if you have a positive frame of mind. Regardless of how negative and furious the customer gets, it is important that you do not let it affect your mood.

No excuses: No matter what the problem is, or who you think caused it, do not blame other people or try to make excuses. Instead, take the initiative and do whatever you can to solve the problem.

R **3** Find the synonyms in the text for the following words and expressions:

1. to stop sth. from happening
2. disagreement
3. to aim
4. to fix
5. attitude
6. very angry
7. reason
8. to take control

R **4** KMK II Sie lesen den oben stehenden Artikel im Internet. Beantworten Sie die folgenden Fragen dazu in ganzen Sätzen auf Deutsch.

1. Warum ist es wichtig zu wissen, wie man mit unzufriedenen Kunden umgeht?
2. Was sollte man tun, wenn der Kunde sehr verärgert ist?
3. Warum ist es wichtig, dem Kunden ganz genau zuzuhören?
4. Wie kann man einem Kunden zeigen, dass man sein Problem ernst nimmt?
5. Wann ist es gerechtfertigt, jemand anderem die Schuld für das vorliegende Problem zu geben?

I **5** Choose one of the problems you listed in exercise A1 and create a short role play. One of you should play the unhappy / angry customer and the other the employee. Act out your role play in front of the class.

WORD BANK

instruction manual – Bedienungsanleitung
plumbing and heating – Heizung und Sanitär
underfloor heating – Fußbodenheizung

09

NACH DIESEM MODUL:

- ✓ kann ich mich zu den wichtigsten aktiven und passiven Sicherheitssystemen eines Autos auf Englisch äußern,
- ✓ kann ich einige Innovationen im Bereich Automobilsicherheit auf Englisch benennen.

1

2

3

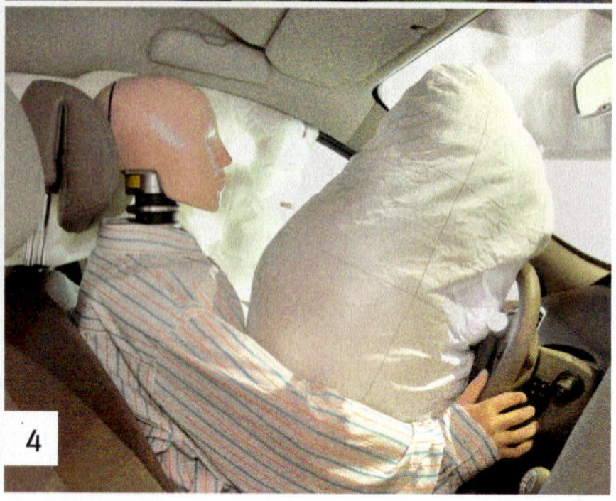
4

Safety

As customer demand for safer cars rises, more and more life-saving technologies are being built into cars in order to reduce fatalities. The term 'active safety features' is used to talk about those features in the car that help the drivers avoid accidents. The term 'passive safety' refers to the safety features that protect the occupants in the event of a crash.

1 Discuss the pictures above and answer these questions.

1. Which safety features are shown in pictures 1, 2 and 4? Are they active or passive safety features?
2. What safety feature(s) would be useful in picture 3?
3. Which of the safety features shown do you think is most important?
4. Are all of these safety features standard in cars today?

2 Find three passive safety systems in the list below.

anti-lock braking system • deformation zones • electronic stability control • intelligent speed adaptation • brake assist system • traction control system • seat belts • collision warning/avoidance • adaptive or autonomous cruise control system • airbags

P **3** What other safety features do you know? Check the Internet and prepare a short presentation on one active and one passive safety feature.

→ **PHRASES:** Presentations

A | Safety in SUVs

1 Find out what the acronyms stand for. Which of them do not refer to safety features?

ACC • ESP • ABS • TC • ASR • ARS • BHP • FWD • LHD • TPMS • ICRS

I **2** What are the differences between an SUV and a normal estate car? Brainstorm in class and collect your ideas on the board. What effects could these have on any safety issues and what do car designers have to do to deal with the challenges?

R **3** Read the text about how designers use different safety features to make SUVs safer. Tick the ideas you collected in exercise 2 that are also mentioned in the text.

SUV safety aspects

One difference between an SUV and a normal estate car is its height. That was one of the features that made the car such a success despite the fact that it was not very fuel-efficient. Drivers like the elevated seating position and so do their children! It became the ideal family car.

However, there is a problem with tall vehicles. The centre of gravity is also higher and that means they can tip over more easily; especially if the driver suddenly has to swerve to avoid hitting an obstacle in the road. To compensate this, car designers included a lot of extra active and passive safety features to ensure that the vehicle is safe and prevent it from rolling over. They reinforced the roof structure with ultra high-strength steel and made the frame structure and side panel stiffer. They also designed a new roll-over prevention system where the electronic stability program uses a gyro sensor to monitor the speed and angle of the car. If the car reaches its terminal angle (that is the point at which it will roll over), the stability control and anti-skid traction control system are activated. This then reduces the engine power and brakes each of the wheels individually so that the vehicle can still be steered.

4 Find synonyms in the text for the following words.

1. triumph
2. raised
3. perfect
4. change direction
5. to bypass
6. barrier
7. strengthened
8. to check
9. slip
10. guided

5 **UNDERSTANDING TECHNICAL ENGLISH**
The following words have different meanings. Make sentences that illustrate their technical and everyday meanings.

1. estate 2. tip 3. monitor 4. roll 5. terminal

R **6** Listen to an interview about active safety features of the SUV. Take notes on the active safety features that are mentioned.
◎ A 4.13

R **7** Listen again and say what scenarios are mentioned that will inevitably cause an SUV to roll over? What happens after an accident?
◎ A 4.13

WORD BANK

anti-skid traction control – Traktionskontrolle und Anti-schleudersystem
gyro sensor – Kreiselsensor
occupant – (Fahrzeug-)Insasse
safety features – Sicherheitsaus-stattung
side panel – Seitenverkleidung
steer (v) – lenken, steuern
ultra high-strength steel – ultra-hochfester Stahl

WORD BANK ◎

seat belt tensioner – Gurtstraffer
deploy (v) – auslösen
B-pillar – B-Säule
hazard warning light – Warn-blinklicht
side curtain – Seitenairbag

B | Safety features

R **1** **Match the sentence parts to explain the functions of the different safety features.**

1. Active headlights
2. An active headrest
3. Blind-spot detection
4. Lane-departure warning
5. Pre-crash sensing
6. Seat belt force limiter
7. Two-stage airbags
8. Tyre pressure sensors
9. Run flat tyres
10. Driver drowsiness sensor
11. Adaptive radar cruise control
12. Seat-occupant sensing

a. rises to support the occupant's head when the seatback is moved backward in a collision to help prevent whiplash.
b. monitor the pressure of the tyres.
c. detects an impending collision using speed and brake sensors, raises seatbacks, tightens seat belts and closes the sunroof.
d. minimises chest injuries from the belt by allowing a small amount of slack back into the seat belt.
e. allow the driver to illuminate around curves.
f. alarm is triggered if the vehicle drifts across side or centre markers.
g. sensors detect vehicles in the adjoining lanes and warn the driver.
h. detects typical indicators of when a driver might fall asleep and sets off an audible alarm or vibration warning.
i. uses radar sensors in the radiator grille to detect the speed and distance of the vehicle ahead.
j. senses when the seat is empty or occupied by a smaller person and turns off or reduces the speed at which the airbags are deployed.
k. have extra-rigid sidewalls to still enable the vehicle to be driven when a tyre loses pressure.
l. only deploy the first stage for smaller occupants or in low-speed crashes.

P **2** **Which safety features would you choose to have in your next car and why?**

3 **Complete the text about the basics of airbag inflation using the words in the box.**

crash sensor • inflator • steering wheel • inflated • gas • spark

The airbag is folded and stored safely in the **1**. The **2** is an inflating device filled with sodium azide (NaN_3) and it is connected to a **3**. When this sensor receives the information that a crash has happened, it triggers the inflator and a **4** lights the sodium azide and very quickly nitrogen **5** is produced. The airbag is deployed and is inflated with the hot nitrogen. So that the driver can move, the bag does not stay **6** for long. The gas escapes through the very fine holes of the nylon material it is made of.

P/I **4** **Make a drawing that explains how an airbag works.**

C | Testing safety features

M **1** KMK II

Ihr Vorgesetzter hat Ihnen diesen Zeitungsartikel gegeben und Sie gebeten, ihn auf Deutsch zusammenzufassen. Er möchte auf dieser Grundlage die folgenden Fragen in einem Vortrag beantworten können.

- Wie hat sich die Sicherheitstechnik in den letzten 20 Jahren verändert?
- Müssen sämtliche Sicherheitssysteme in alle PKWs eingebaut werden?
- In welchen Autos werden die neuen Sicherheitssysteme zuerst eingebaut?
- Warum sind manche Kunden nicht bereit, für Sicherheitssysteme zu bezahlen?
- Was ist EuroNCAP?
- Wo kann man herausfinden, welche Autos am sichersten sind?
- Wie werden sich die Sicherheitssysteme in Zukunft weiterentwickeln?

WORD BANK

inflated – aufgeblasen
nitrogen – Stickstoff
sodium azide – Natriumazid (NaN_3)
whiplash – Schleudertrauma

Building and testing safety features in vehicles

For several years more and more car manufacturers have focused on developing strategies to help drivers maintain control of their vehicle in order to prevent accidents.

At the beginning, safety technologies were developed to protect the driver and passengers inside the vehicle. Later, designers concentrated on being able to see everything around the outside of the car in order to prevent accidents. They also thought that if a crash could not be avoided, anyone inside the car should at least be able to survive with as few injuries as possible. However, car designers today not only have to consider the safety of the occupants, they also have to think of ways to minimise injuries to pedestrians.
Some safety features, such as seat belts and occupant sensing technology, are now legal requirements, but others are still optional. Moreover, some safety features can be very expensive so a lot of drivers can either not afford them or are not convinced they are necessary. This means it is nearly always the luxury vehicles that have them first. Nevertheless, vehicle safety rating systems such as NCAP (New Car Assessment Program) are helping to set new standards. Assessment programmes like EuroNCAP provide consumers with a realistic, independent rating of how the safety features of some of the most popular cars sold in Europe perform. They have become a driving force for significant safety improvements to new car design. The results of all the tests are published so customers are more aware of what safety features exist. Before buying a new car, they

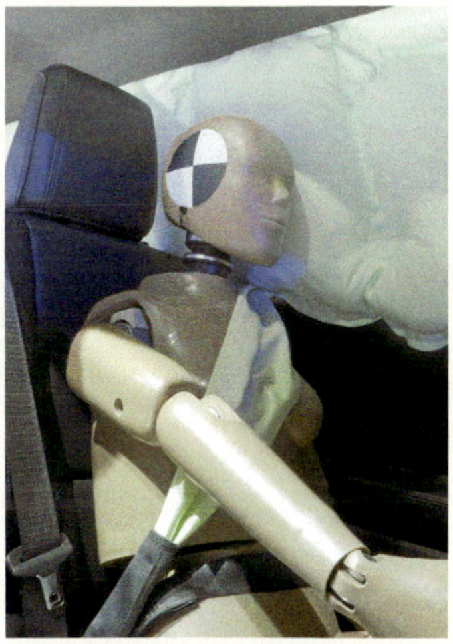

can use these test results to decide which is the right car for them.
Many people believe that in the future all the safety technologies in different vehicles will be integrated and linked so that vehicles can communicate with each other. The idea is that a car would receive a command to steer itself away from a dangerous situation even if the driver didn't react.

P **2**

Search the Internet and find out what the latest pedestrian safety regulations are and what car manufacturers are doing to meet those new regulations. How safe is your dream car? Write a review.

10

NACH DIESEM MODUL:

- ✓ kenne ich die englischen Begriffe für die wichtigsten Bauteile im Bereich Räder und Bremsen,
- ✓ kann ich einfache Arbeiten an Bremssystemen auf Englisch erklären,
- ✓ kann ich Standardsituationen in der Werkstatt auf Englisch bewältigen.

1

2

3

4

Wheels and brakes

Doing an apprenticeship as a motor vehicle mechatronics technician or working at a car garage as a service mechanic means dealing with different systems such as the power unit, the drivetrain, the electrical system and car suspension. The brakes, the wheels and the tyres are all linked to the suspension system. There is a lot of wear and tear on brakes and wheels, which is why car mechanics spend a lot of their time repairing or replacing them. Modern car wheels, especially light alloy wheels, are often just called 'rims' although the rim of a wheel is really only the outer edge. 'Wheel' is now also a common term used for the complete wheel, including rim and tyre.

1 Describe the pictures above. What is the difference between picture 1 and 2? Have you seen the system shown in picture 3 before? What system can you see in picture 4?

M **2** Write a short summary of the text above in German.

P **3** Write a few sentences about the braking system of a car.

4 **UNDERSTANDING TECHNICAL ENGLISH**
The following words are used differently in technical and everyday situations. Make sentences to show their different meanings.

1. rim 2. tear 3. tyre (BE)/tire (AE) 4. wear

A | Wheels and tyres

1 Changing the wheels of a car is one of the first tasks trainee mechanics are given. However, before they start working on a vehicle, they need to know some general rules. Match the German rules to their English equivalents.

1. Sichern Sie zuerst das Fahrzeug mit Keilen gegen das Wegrollen.	a. The base of the jack must always be positioned directly below the jacking point.
2. Das Lenkrad sollte in Geradeaus-Stellung eingerastet sein.	b. Light alloy wheels and steel wheels sometimes have different wheel bolts.
3. Der Wagenheber darf nur an den jeweiligen Wagenheber-Ansatzpunkten angesetzt werden.	c. First secure the vehicle with chocks to prevent it from rolling away.
4. Der Fuß des Wagenhebers muss immer direkt unter dem Ansatzpunkt positioniert werden.	d. Always use the same type of tyres and with the same tread on one axle.
5. Radschrauben müssen über Kreuz mit dem vom Hersteller vorgegebenen Anziehdrehmoment angezogen werden.	e. The jack must only be positioned on the appropriate jacking point.
6. Nur die vom Hersteller mitgelieferten Radschrauben dürfen benutzt werden.	f. Check the tyre pressure.
7. Leichtmetall- und Stahlräder haben manchmal unterschiedliche Radschrauben.	g. Wheel bolts must be tightened crosswise to the tightening torque defined by the manufacturer.
8. Auf einer Achse immer Reifen gleicher Bauart und gleichen Profils verwenden.	h. Only the wheel bolts supplied by the manufacturer may be used.
9. Prüfen Sie den Reifendruck.	i. The steering wheel should be locked in the straight ahead position.

2 Look at exercise 1 again and find the English terms for the German words below.

Keil • Wagenheber • Wagenheber-Ansatzpunkt • Radschraube • anziehen • Drehmoment • Leichtmetallrad • (Reifen-)Profil

P **3** Write down four more rules you know to do with changing the wheels on a vehicle. Present them to the class.

R **4** ◎ A 4.14 Listen to the dialogue and answer the questions in complete sentences.

1. In which situation are all-season tyres useless?
2. Why is it difficult to brake with all-season tyres in winter?
3. Give two reasons why snow tyres are safer to drive at low temperatures.
4. What reasons are given for not driving with snow tyres in summer?
5. Why is it advantageous to have two sets of wheels?

I **5** Use the role cards on pages 104 and 105 to role-play a sales talk in which you help a customer to buy new tyres for their cars.

WORD BANK

brake – Bremse
light alloy wheel – Leichtmetallrad, Alufelge
rim – Felge, Radkranz
suspension – Federung, Aufhängung
tread – (Reifen-)Profil
tyre (BE)/tire (AE) – Reifen
wear and tear – Verschleiß
wheel – Rad, Felge

WORD BANK ◎

all-season tyre – Ganzjahresreifen
braking distance – Bremsweg
grip – Haftung
rubber compound – Gummimischung
snow tyre – Winterreifen

→ **PHRASES:** Dealing with customers

B | Disc brakes and drum brakes

Two important brake types are drum brakes and disc brakes. Some vehicles have disc brakes on all four wheels while others have disc brakes on the front wheels and drum brakes on the rear.

1 Match each term to its function.

1. brake caliper	a. carries the lining of the drum brake
2. brake booster	b. is where the brake fluid is compressed by a piston
3. brake shoe	c. pushes the pads against the disc to slow the car down
4. brake cylinder	d. the thing you press with your foot to slow the car down
5. brake pedal	e. multiplies the force applied to the master cylinder

2 Which of these drawings shows a disc brake and which shows a drum brake? Label the different parts in pictures 1 and 2 with the words in the boxes.

Picture 1: brake disc • brake pad • caliper • wheel hub • wheel stud • piston

Picture 2: brake drum • brake lining • brake shoe • piston • return spring • wheel stud • wheel cylinder

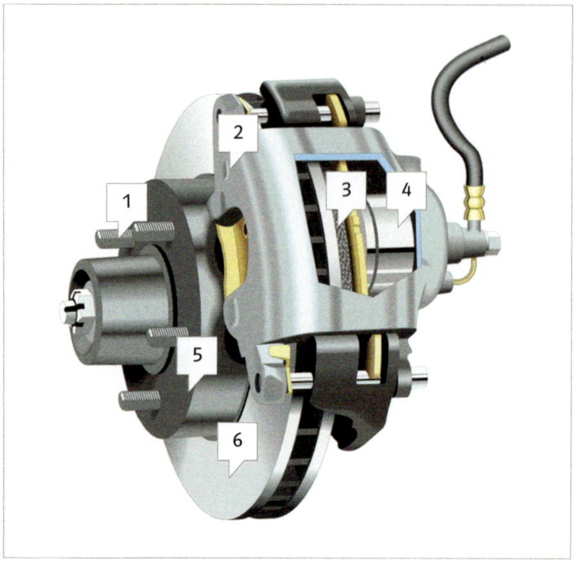

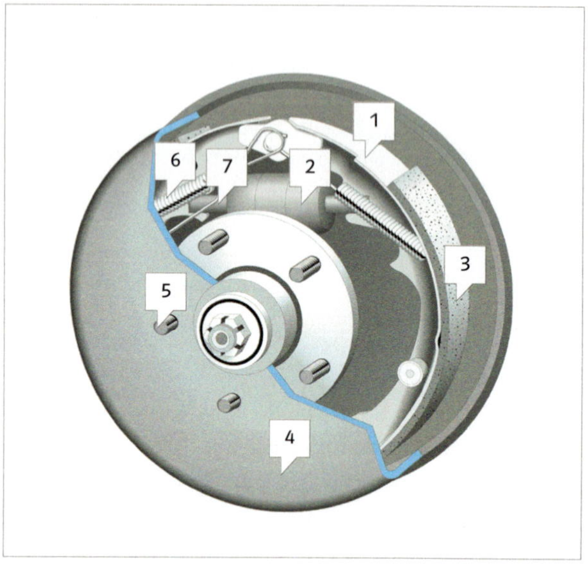

3 Look at the drawings again and describe the differences between the drum brake and the disc brake.

P **4** Search the Internet and find out how drum brakes and disc brakes work. Write a summary comparing the two brake types.

C | Braking systems

Brakes are used to slow a car down, to stop it completely or to hold it in a certain position – on an incline for example. Passenger cars are equipped with a hydraulic brake system whereas trucks and buses have an air brake system.

1 Use the words from the box to describe the following drawing of a brake system.

brake booster • brake drum • return spring • wheel-brake cylinder • brake disc • brake pressure reducer • brake pedal • brake circuit • tandem master cylinder • brake shoe • brake cylinder

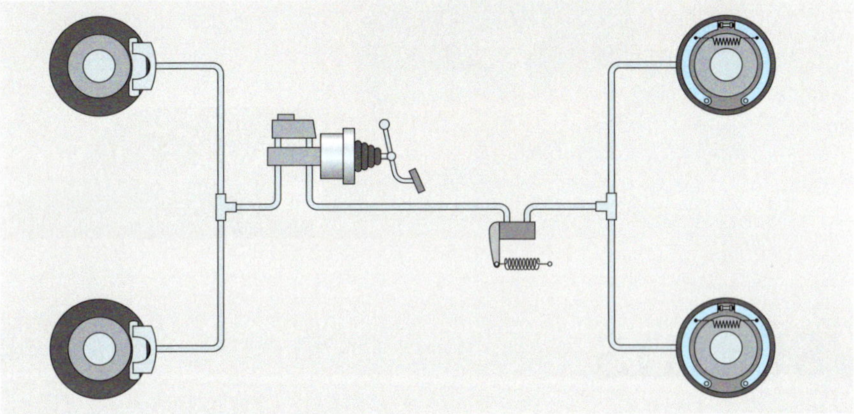

M **2** KMK II Sie haben im Internet folgende Montageanweisung zu Bremsscheiben gefunden. Übertragen Sie die Montageanweisung in einen deutschen Fließtext, der als Teil einer Infomappe an neue Auszubildende in Ihrer Firma weitergeleitet werden soll.

ww#.brake_discs_info.com

Instructions for changing brake discs

→ First fasten the handbrake of the vehicle.
→ Loosen the lug nuts on the wheel with a wheel wrench.
→ Jack up the vehicle.
→ Remove the wheel.
→ Compress the piston back into the brake caliper with the c-clamp.
→ Take out the bolts that hold the caliper.
→ Remove the caliper and hang it on the shock absorber.
→ Next remove brake pads and the caliper bracket.
→ Slide off the old rotor.
→ Clean off all the protective oil coating from the new rotor.
→ Mount the new rotor.
→ Put back the caliper bracket and check the torque with a torque wrench.
→ Replace the brake pads and slip the caliper back over the rotor.
→ Replace the wheel and tighten the lug nuts.
→ Finally lower the car again, pump the brake pedal and you are finished.

WORD BANK

bracket – Klammer
brake booster – Bremsverstärker
brake caliper – Bremssattel
brake fluid – Bremsflüssigkeit
brake shoe – Bremsbacke, Bremsklotz
c-clamp – Schraubzwinge
lining – *(hier:)* Bremsbelag
lug nut – Radmutter
pad – Klotz, Belag
piston – Kolben
return spring – Rückstellfeder
stud – Spike
torque wrench – Drehmomentschlüssel
wheel hub – Radnabe
wheel wrench – Kreuzschlüssel

11

NACH DIESEM MODUL:

- √ kann ich einfache Erklärungen zum Kühlsystem auf Englisch geben,
- √ weiß ich, wie die wichtigsten Teile der Abgasanlage auf Englisch heißen,
- √ kann ich mich zu einigen Problemen mit Abgasanlagen auf Englisch äußern.

1

2

3

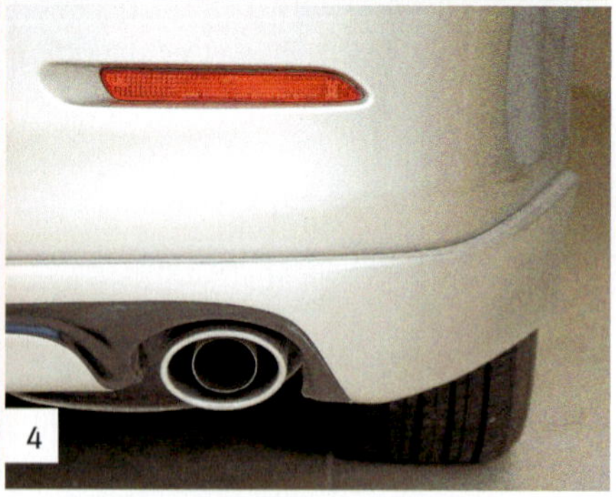
4

Car systems

Two car systems are the cooling system and the exhaust system. The cooling system stops the engine from overheating. An engine produces a lot of heat when it burns fuel and serious damage is caused if it overheats. However, an engine runs best and lasts longer when it runs at a fairly hot temperature. So another important job of the cooling system is to heat the engine as quickly as possible and keep it running at a constant temperature. There are two types of cooling systems: air-cooled and liquid-cooled. The cooling systems found in cars today are liquid-cooled. Some of the heat that is produced when fuel is burnt leaves the vehicle in the form of gas via the exhaust system. These gases are harmful to the environment. The exhaust system makes them less harmful and also reduces the noise of the engine.

1 Name all the parts of the car you can see in the pictures above.

I **2** Look at the large part in the centre of the front of the car in pictures 1 and 2. What do you think the vents are for and what lies directly behind those vents?

P **3** Make a list of all the parts you know that belong to the exhaust system. Present your results in class.

P **4** Search the Internet to find out what can happen when a car engine overheats. Write a short descripton in your own words.

A | The liquid cooling system

1 Match the words on the left to the correct explanations on the right.

1. antifreeze	a. ignites the petrol/air mixture to help start the engine
2. fluid	b. liquid used to stop the engine from overheating
3. coolant	c. flat tank where hot liquid is cooled as it passes through
4. fan	d. flexible tube often made of plastic
5. heater core	e. opens and closes to control the way fluids or gases flow
6. hose	f. synonym for 'liquid'
7. pressure cap	g. substance with low freezing and high boiling point
8. radiator	h. blades that rotate to reduce temperature using air
9. radiator drain cock	i. measures the temperature of the coolant
10. spark plug	j. a small kind of radiator that uses hot coolant from the car engine to heat the car interior
11. valve	k. a bolt at the bottom of the radiator that is removed to let all the coolant out of the radiator
12. thermostat	l. sends cold coolant through hoses from the radiator of the engine
13. water pump	m. radiator lid that keeps the pressure level so that the coolant boils at a higher temperature

R **2** ◎ A 4.15 Listen to an engineer explaining a basic cooling system. Write down all the words from exercise 1 that you **do not** hear.

R **3** ◎ A 4.15 Listen again and complete the sentences below with the correct expressions from exercise 1.

1. The ? is the most important part of the cooling system.
2. The ? rotate when the engine is running.
3. The ? is attached to the radiator via the lower radiator hose.
4. The valve of the ? opens so the hot coolant can flow back into the radiator and be cooled again.
5. You need the ? to heat the inside of the car.

R **4** ◎ A 4.15 Listen again and answer the questions below in full sentences.

1. What does the engineer say the radiator looks like?
2. Where in the car is the radiator positioned?
3. What does coolant normally consist of?
4. Why do the fans rotate?
5. When does the valve of the thermostat open?
6. What happens when the pressure in the radiator gets too high?

I **5** Use all the information from the above exercises to write down and present your own description of a cooling system to the class.

WORD BANK

antifreeze – Frostschutzmittel
blade – Schraubenblatt, Flügel
coolant – Kühlmittel
cooling system – Kühlsystem
exhaust system – Abgassystem
fan – Ventilator, Gebläse, Lüfter
heater core – Wärmetauscher
hose – Schlauch
ignite (v) – zünden
liquid – Flüssigkeit
overheating – Überhitzung
pressure cap – Druckverschluss
radiator – Kühler
radiator drain cock – Kühlerablasshahn
spark plug – Zündkerze
valve – Ventil
vent – Lüftungsöffnung

WORD BANK ◎

grille – (Kühler-)Grill
release (v) – freisetzen, entlassen

B | The exhaust system

R **1** Read the article from a car magazine and say what you can see in the pictures.

1

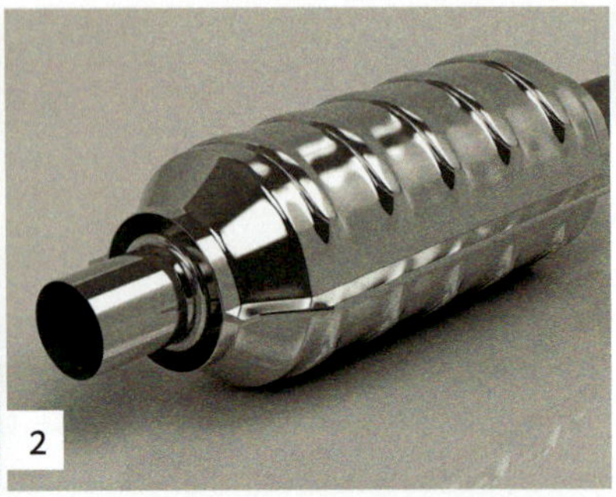
2

3

What is the exhaust system for and how does it work?

The exhaust system of a car not only takes the gases directly from the engine and transports them through the exhaust pipe and out through the tail pipe. It also makes the car more environmentally friendly because it reduces the toxic waste as well as the noise that comes from the combustion in the engine. It also helps to ensure that the engine uses less fuel.

The exhaust system is often referred to as the exhaust pipe, but there is more to it than just a pipe. The main parts of an exhaust system are the exhaust manifold, the catalytic converter, the silencer and the tail pipe. Other parts are the resonator, the oxygen sensor, brackets, gaskets, clamps and hangers.

The exhaust manifold is a kind of large pipe that has several smaller pipes. Each of these small pipes is attached to a cylinder and takes in hot air or gases from the combustion chamber of the engine. They also take in any fuel that has not burnt completely. This is then burnt in the manifold. The exhaust gases then go through the exhaust pipe and into the catalytic converter. The catalytic converter continues to burn the fuel and uses oxygen to help change carbon monoxide and hydrocarbons into water and carbon dioxide. Nitrogen oxide levels can also be reduced and so the exhaust emissions become both safer and more environmentally friendly.

When the exhaust leaves the converter, it goes into another connecting pipe which takes it to the silencer. This is also called the 'muffler'. It 'muffles', that is it reduces, the sound in the exhaust system. The exhaust fumes are then transported to the tail pipe and emitted into the environment. They are now a lot cleaner and safer.

R **2** Read the article again and answer the questions below in complete sentences.

1. What tasks are carried out by the exhaust system?
2. According to the article, which part of the exhaust system takes the heat and gases directly from the engine?
3. How does the exhaust system make the emissions more environmentally friendly and less harmful?
4. Which part of the car makes the noise of the engine quieter?

3 **UNDERSTANDING TECHNICAL ENGLISH**

The following words are frequently used in car production. Translate both their technical and everday meanings into German. You can use a dictionary.

1. fan 2. core 3. cap 4. radiator

C | Dealing with problems in the systems

M/I **1** Jonas is a German apprentice who is doing a work placement in a garage in Ireland. He is talking to his Irish colleague Patrick about the kind of repairs he has done to the exhaust system in Germany. Read what they talk about and make a dialogue in English. Be prepared to act it out in class.

Patrick	Jonas
Er will wissen was für Erfahrungen Jonas mit Reparaturen an Abgasanlagen hat. →	Er sagt, dass Korrosion das größte Problem war; Auspuffrohre bekommen dann Rostlöcher.
Fragt wie man das merkt. →	Ein Kunde kam in die Werkstatt, weil sein Auto plötzlich sehr laut wurde – wie ein Rennwagen. Der Schalldämpfer war durchgerostet *(to rust through)*.
Will wissen, ob Jonas den Schalldämpfer reparieren konnte. →	Nein, das geht fast nie. Sie haben den Dämpfer ersetzt.
Erkundigt sich, was es sonst noch für Probleme gab. →	Ein Mann hat nicht verstanden, warum er plötzlich mehr Benzin verbrauchte. Wir haben dann herausgefunden, dass es an der Lambdasonde lag.
Fragt, was dann gemacht wurde. →	Die Sonde musste ersetzt werden.
Gab es sonst noch was? →	Oft sind es nur die Aufhänger, die abbrechen *(break off)*. Dann ist der Auspuff locker *(loose)* und das hört man auch.

R **2** Read the sentences below and put them in the correct order.

a. Replace plug and open radiator cap. Careful – only when cold!
b. Fill in the flush solution and add water until the radiator is full.
c. Open the radiator plug and drain all the coolant out into the bucket.
d. Turn engine off and wait until engine has cooled down again completely.
e. Refill radiator with coolant/water mix using a funnel and close radiator cap.
f. Close radiator cap, start engine and wait for 10 minutes.
g. Remove plug and drain all the cleaning solution out of the radiator.
h. First make sure the engine is cold and place a bucket under the radiator.
i. Put the plug back and open radiator cap.

WORD BANK

bracket/clamp – Klammer
catalytic converter – Katalysator
cleaning solution – Reinigungslösung
combustion chamber – Brennkammer, Verbrennungsraum
exhaust manifold – Abgaskrümmer
flush solution – Spüllösung
gasket – Dichtungsring
hydrocarbon – Kohlenwasserstoff
muffler/silencer – Schalldämpfer
nitrogen oxide – Stickstoffoxid
oxygen sensor – Sauerstoffsensor, Lambdasonde
tail pipe – Endrohr

R **3** KMK II ⊙ A 4.16 Ihr Vorgesetzter gibt Ihnen einen Podcast mit Informationen auf Englisch über Probleme, die im Kühlsystem auftreten können. Er bittet Sie, sich den Podcast anzuhören und die Symptome, Ursachen und Problemlösungen übersichtlich auf Deutsch für die neuen Azubis aufzulisten.

WORD BANK ⊙

crack – Riss
leak (v) – tropfen, lecken

12

NACH DIESEM MODUL:

- √ kann ich auf Englisch erklären, was bei einem Kundendienst gemacht wird,
- √ kann ich einfache Reparaturen an einem Auto auf Englisch erklären,
- √ kann ich mich auf Englisch zu TÜV-Bestimmungen äußern

Maintenance and repairs

Making sure your car remains a safe vehicle means you have to have it serviced at regular intervals. When you buy your car, you will also be given a service booklet that states when the car was delivered and warranty begins. It contains all the details of your vehicle, a schedule of when it has to be serviced and what work is carried out in each service. When the service is finished, make sure it is entered in the booklet. Some car manufacturers will ask to see a complete service record if you have a problem during the warranty period and you wish to make a claim. It is a good idea to keep the service booklet in the glove compartment and when you sell the car, make sure you pass it on to the buyer. After all, it proves how well you looked after your car.

1 Read the text and answer the following questions.

1. Why is it necessary to have your car serviced regularly?
2. Why is each service recorded?
3. In which way might good maintenance be linked to selling your car?

2 Describe what is happening in each picture. Tell your partner which of these tasks you have carried out and how you did them.

3 Make a list of more tasks that are performed when a car is serviced.

A | Having your car serviced

1 Complete the text using the words from the box.

display • oil • consumption • maintenance • breakdown • replacing

Car **1** can be expensive but if you look after your car, it can prevent you having a **2** or, even worse, causing an accident and injuring yourself or others. In addition to that it will keep the engine running smoothly, help reduce your fuel **3** and maintain the value of your car. So in the long run it may even save you money.
Not every car service is the same. Some things need checking or **4** more often than others. For example an **5** service is normally carried out every 12 months or every 15,000 km, and a full car inspection every 24 months or every 30,000 km. Your car **6** will tell you when the next oil service is necessary so it is important to make sure it is reset after every service. You can find out more information in your service booklet.

R **2** Put the jumbled sentence parts in the correct order.

1. should be changed • Engine oil • replaced • and the oil filter
2. must be • 24 months • The brake fluid • every • changed
3. checked • brake pad thickness • to be • needs • The
4. for • All engine components • leaks • are checked • and damage
5. wash/wipe system • Function • windscreen • is checked • of • and • headlight
6. adjusting • Spray • and • wiper blades • nozzles • need
7. to • on • for • front axle • checked • have • play • and • be • Joints • looseness

3 SAFETY FIRST ⚠

1. Match the warning symbols in the picture to the following messages. One message does not fit.

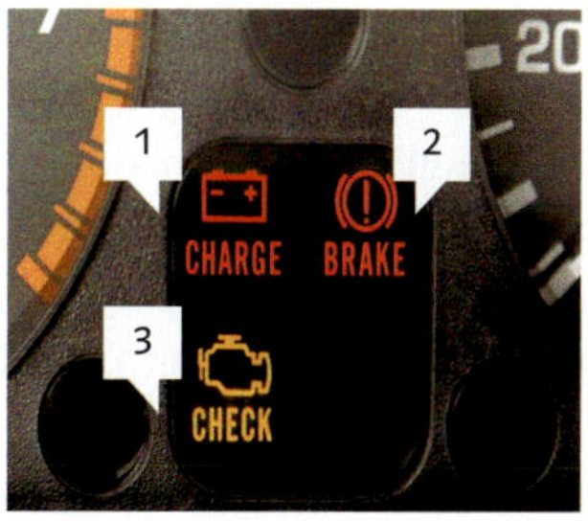

a. Check the engine!
b. Low battery – visit workshop!
c. Problem with the braking system!
d. Fasten seat belts!

2. Find a possible cause for each warning message from the list below.

- The oxygen sensor might be damaged.
- One or more passengers have not fastened their seat belts.
- The light was not switched off over night.
- There is not enough brake fluid in the reservoir.

P **4** Brainstorm what other warning symbols you know and when they light up. Present your ideas to the class.

WORD BANK

joint – Verbindung
nozzle – Spritzdüse
oxygen sensor – Sauerstoffsensor, Lambdasonde
wash / wipe system – Scheiben-wasch- / wischanlage
wiper blade – Scheibenwischerblatt

B | Do it yourself

Having even the smallest car repair done in a garage can be very expensive. So if a car is older and no longer has fully comprehensive insurance, owners may like to do some small repairs themselves.

1 Describe the damage shown in the pictures using some of the the words the box.

scratch(es) • damaged/dented hood/fender • flat tire • dents • damaged side/rear/front panel • missing lights • smashed headlights • rust • broken windshield

P **2** Use the words in exercise 1 and a dictionary to describe other possible car damage. Make a list in English and German. Compare your lists in class.

P **3** The jumbled instructions below explain how you can remove scratches on the paintwork in 10 steps. Put the instructions in the correct order.

a. Cover a sanding block with fine sandpaper and dip it into the soapy water.
b. Finally, use a soft cloth to polish the car with automotive wax – this will protect it and make it shine.
c. First wash the scratched area and wipe it so you can see the damage clearly.
d. If necessary spray the area lightly with clear coat.
e. Let clear coat dry for 15 minutes before you spray on a second clear coat.
f. Put a couple of drops of washing up liquid into a bowl of cold water.
g. Sand carefully until no more scratches are visible.
h. Sand the damaged area softly and rinse the sandpaper regularly.
i. You can remove any fine lines that may appear during sanding by rubbing a bit of polishing compound with a soft sponge over the area until the compound disappears.
j. Wash the area again to remove any dust from the sanding process.

M **4** Check your answers with a partner and explain any differences. Then describe the steps to each other in German.

5 Use your lists from exercise 2 and make a list of car repairs that cannot be done by amateurs.

P/I **6** Group A: Search the Internet and find out how to repair a dent in the car. Prepare a presentation explaining the steps in English to the class.
Group B: Search the Internet and find out how to change the engine oil of a car. Prepare a presentation explaining the steps in English to the class.

→ **PHRASES:** Presentations

C | MOT

In many countries you are required by law to have your car tested by the MOT (Ministry of Transport) after certain specified intervals. In this test the safety, roadworthiness and the exhaust emissions are checked. If your vehicle meets the required standards, you will receive a certificate and a disc. This disc is displayed in the car and says when the car has to be tested again. If your vehicle fails the test, you have to have it repaired and then tested again, so it is a good idea to take it to a garage and have a full inspection done before you drive to an MOT centre.

R **1** ⊙ A 4.17 **Ms Hall's car needs its first MOT. She phones up the garage to ask for an appointment. Listen and answer the following questions.**

1. What is Ms Hall's car registration number?
2. How old is Ms Hall's car?
3. Where is the MOT normally carried out?
4. What service will the garage do?
5. When do they make the appointment for?
6. Will Ms Hall have to pay for a car while her own car is in the garage?

P **2** **What about the MOT test in your country? Answer the questions below and prepare a one-minute presentation.**

1. What is the test called?
2. When does a car have to have its first MOT?
3. How often do you need to have your car tested?
4. What tests are carried out in the MOT?

M **3** KMK II **Eine Kollegin wird für zwei Jahre nach England versetzt und hat Sie gebeten, ihr die folgenden Informationen über die englischen TÜV-Bestimmungen zu erklären. Lesen Sie den Artikel und schreiben Sie ihr eine deutsche E-Mail mit einer kurzen Textzusammenfassung.**

MOT test

In the UK when a car is three years old, it needs to have its first MOT test. Any car that is older than three years has to be tested every year. It is a legal requirement to try to help save lives, make the roads safer, stop crime and protect the environment.

VEHICLE TESTING STATION
APPROVED BY
THE DEPARTMENT OF TRANSPORT

It is important to know that the MOT is not a car service and if your car passes the test it does not mean it will run perfectly until the next MOT is performed.

Last year changes were made to the MOT test to be in line with the minimum European standards. These changes are mostly to do with the many complex electronic and electrical features modern cars now have. The new test includes making sure the fuel system does not leak and the fuel cap fits properly. The exhaust emissions are tested and the exhaust pipe must not have holes or be noisy. The horn, lights and windscreen wipers have to work properly and seats have to be able to move forwards and backwards. Additionally the steering and suspension systems are checked and all the restraint systems and the warning lights for the airbags must function correctly. All the doors have to open and close properly, the door locks and the latch of the bonnet must close securely. The test also ensures that the condition of the brakes meets all the requirements and all the warning lights for ABS, ESP, brake fluid, tyre pressure and parking brake, etc. work properly.

WORD BANK

dent – Delle, Beule
exhaust emissions – Abgase
exhaust pipe – Auspuffrohr
fender (AE)/wing (BE) – Kotflügel
fuel cap – Tankdeckel
hood (AE)/bonnet (BE) – Motorhaube
MOT – (britischer) TÜV
panel – Blech
polishing compound – Polierpaste
restraint system – Rückhaltesystem
roadworthiness – Verkehrstauglichkeit
scratch – Kratzer
suspension – Federung, Aufhängung
windshield (AE)/windscreen (BE) – Windschutzscheibe

WORD BANK ⊙

car registration number – Autokennzeichen
courtesy car – Leihwagen

→ **PHRASES:** Written communication

13

NACH DIESEM MODUL:

- √ kann ich mich zu einigen LKW-Typen auf Englisch äußern,
- √ kann ich Eigenschaften und Merkmale von LKWs auf Englisch beschreiben,
- √ kann ich die Maße von LKWs auf Englisch angeben.

1

2

3

4

Trucks and lorries

In British English the word 'lorry' is traditionally used to refer to LGVs (large goods vehicles) or HGVs (heavy goods vehicles) and the word 'truck' is often used for smaller and lighter transporters. However, in the USA the word 'truck' can refer to any vehicle used to transport both large and small amounts of all types of goods. Buses are known as PCVs (passenger carrying vehicles). The term 'commercial vehicles' is often used for all types of vehicles that are used for commercial reasons.

1 Match the vehicles to their description and to the pictures above.

1. concrete mixer truck	a. can be used for long distance transportation
2. tipper truck	b. has a device which mixes sand, gravel and water to form concrete
3. pick-up truck	c. here the rear platform can be raised at the front
4. long-haulage lorry	d. small truck with an enclosed cab and open rear

P **2** Discuss the following questions.

1. Which vehicles in the pictures can you drive off-road?
2. What are the main differences between a passenger car and a truck or lorry?
3. Which vehicles can you find on a construction site?
4. Which vehicle would you use to transport heavy furniture, machines, etc.?
5. What kind of things would you transport in a pick-up truck?

A | Truck categories

1 Match the English expressions to their German translation.

1. commercial vehicle	a. Führerhaus
2. weight category	b. Gewichtsklasse
3. special permit	c. Nutzfahrzeug
4. articulated lorry	d. Sattelschlepper
5. cab	e. Sondergenehmigung

R **2** Read the article below and say if the following statements are true or false. Correct the false statements.

1. All commercial vehicles are built for the same reason.
2. A lorry in the heavy category has to have more than two axles and weighs more than 8 tonnes.
3. A rigid lorry is one where the cab and trailer are permanently attached.
4. It is more difficult to drive an articulated lorry around a tight bend.
5. The inside of a cabin has more features than a passenger car.
6. Safety features are tested in passenger cars before they are built into trucks.

WORD BANK

articulated lorry – Sattelschlepper
driver drowsiness sensor – Fahrerzustandserkennung
GVM (gross vehicle mass) – Bruttogesamtgewicht
HGV (heavy goods vehicle) – Schwerlastwagen
lane-departure warning system – Fahrspurerkennungssystem
LGV (large goods vehicle) – Lastwagen
lorry / truck – LKW, Lastwagen
PCV (passenger carrying vehicle) – Fahrzeug für die Personenbeförderung
pivot joint –Drehgelenk
turning circle – Wendekreis

TRUCK CATEGORIES

Commercial vehicles are built for a special purpose. They are often divided into three main weight categories: **light**, **medium** and **heavy**.

A medium commercial vehicle for example weighs between 4.5 and 8 tonnes and has two axles. In contrast, a lorry with a GVM (gross vehicle mass) of over 8 tonnes that has three or more axles is in the heavy category. Some very heavy lorries are so large or long that they need a special permit and escorts to be able to drive on public roads.

There are rigid lorries and articulated lorries. Articulated lorries are different from rigid lorries in that cab and trailer can be separated. The cab is also called a 'tractor' and is equipped with a pivot joint. This gives the vehicle a smaller turning circle which means the driver can turn the lorry more easily and needs less room.

Lorry drivers can spend a long time in their vehicles and a great deal of attention is paid to the design of the cabin so that all the drivers' requirements can be integrated. Many safety features in passenger vehicles today, such as night vision, driver drowsiness sensor and lane-departure warning system, were first designed for and built into HGVs.

3 Use the terms below to label the cab of a lorry.

a. rear wall	f. rim
b. front panel	g. step
c. A-pillar	h. sunroof
d. roof	i. tyre
e. headlight	j. wheel arch

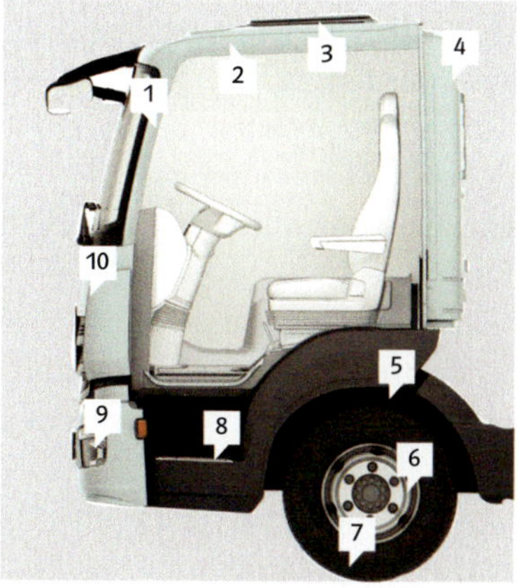

4 Make a list of all the things you would expect to find inside the cabin.

B | Interior cabin features

1 Match the definitions to the words on the right.

1. This is used to keep food fresh and drinks cool.
2. You use this to turn on the lights, for example.
3. This is where warm or cool air comes into the cabin.
4. You need this if you smoke cigarettes.
5. This is where the driver keeps important papers.
6. This is what the driver sits on.
7. This is a flat surface you can place things on.
8. The driver needs this to keep fresh clothes in.
9. The driver uses this when he is tired.
10. The driver will use this when he drinks something while he is driving.
11. You can eat on it and store it away afterwards.
12. The driver leans against it when he is driving.

a. bed (folding mattress)
b. document pockets
c. air vents
d. folding table
e. cup holder
f. refrigerator
g. storage compartment
h. ashtray
i. seat backrest
j. seat cushion
k. switches
l. tray

P **2** Check the Internet and find pictures of cabins that show the features mentioned in exercise 1. Write a description of your favourite picture.

P **3** Imagine you are a truck driver and are going to be away for three weeks. Make a list of everything you will need to have inside the cabin. Present your ideas in class.

4 **UNDERSTANDING TECHNICAL ENGLISH**
What do these frequently used technical abbreviations mean?

HGV • LGV • GPS • PCV • RWD • GVM • PSV

R **5** KMK II A 4.18 John ruft eine englische Fahrschule an, um zu erfahren, wie er eine LKW-Fahrerlaubnis bekommen kann. Hören Sie sich das Gespräch an und entscheiden Sie, ob die folgenden Sätze richtig oder falsch sind. Begründen Sie Ihre Entscheidung auf Deutsch und korrigieren Sie die falschen Sätze.

1. Um einen LKW von 3,7 Tonnen zu fahren, genügt ein normaler Führerschein.
2. Der Test besteht aus zwei Teilen.
3. Im so genannten *hazard perception test* werden die Reaktionen des Fahrers in Gefahrensituationen am Computer geprüft.
4. LKW-Fahrer in Europa müssen alle fünf Jahre ein 40-stündiges Kompetenztraining machen.
5. Als anstehender LKW-Fahrer muss man nachweisen, dass man gesund ist.
6. Um einen LKW mit mehreren Anhängern zu fahren, muss man einen Führerschein für einen Sattelzug seit mindestens einem Jahr besitzen.

R/P **6** A 4.18 You are working for Happy Lorry Driving. Write the email to John Farlow summing up the most important points. If necessary, listen to the telephone conversation again.

→ **PHRASES:** Written communication

C | Truck dimensions

1 You need to know the following expressions to talk about different lorry sizes. Find out the German equivalents for the following terms. Use a dictionary if necessary.

exterior height • sleeper • steering axle • wheelbase • cab interior height • track • tractor • interaxle spacing (2x) • drive (tandem) axle • overall length • kingpin • interior length • swing doors • cab exterior length • exterior width • semi-trailer • effective rear overhang

2 Match the English terms in exercise 1 to the numbers in the pictures below.

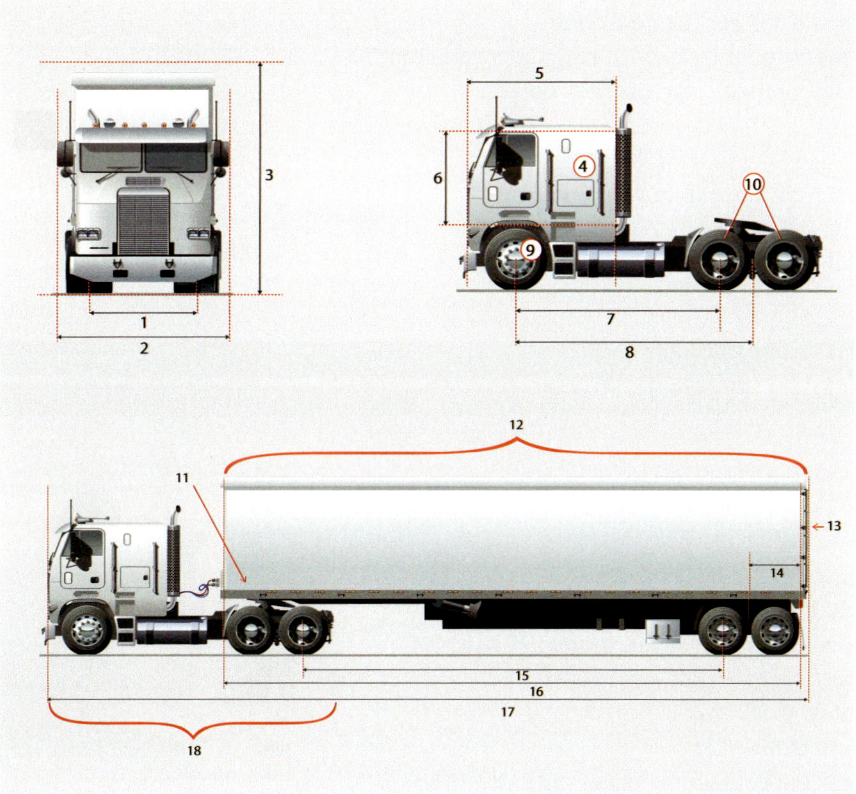

P **3** Look at the pictures again, choose six measurements and write a description for each.

Example:
The overall length of the lorry is measured from the front bumper to the back of the rear bumper.

P **4** Think about your dream lorry and make notes on its features and measurements. Then describe it to your partner and say which routes you would like to drive with it.

WORD BANK

air vent – Luftschlitz
bumper – Stoßstange
kingpin – Zugsattelzapfen
measurements – Maße, Abmessung
track – Spurweite
wheelbase –Radstand

WORD BANK

hazard perception test – Gefahren-wahrnehmungstest

V1 Company tour

lingua tv

Before watching

1. Describe the photo.
2. Who do you think the men in the picture are?

While watching

3. Watch the film and find out the following:
   a. the number of countries SAND Machinery delivers to.
   b. the names of three departments that the men have visited.
   c. the number of people who work for the company.
   d. whether SAND Machinery has other sites and if so, where they are.
4. Watch the film again and answer the questions.
   a. What will Mr Kramer's main tasks be during the internship?
   b. What is made at the company's production site?

After watching

5. Explain the sentences from the clip in your own words.
   a. "I'm eager to get in and get my feet wet."
   b. "... around the globe, around the clock."

V2 Telephoning

lingua tv

Before watching

1. Describe the photo.

While watching

2. What problem does Ralf have at the beginning of the call?
3. Watch the film again and answer the questions.
   a. Who is calling?
   b. Who does he want to speak to?
   c. What is the purpose of his call?
   d. Why can Ralf's boss not contact the caller until the 15th of the month?
4. Find out the following technical information about the vehicle in question:
   a. the engine size
   b. the top speed
   c. the distance that can be travelled on a single charge.

After watching

5. Use the information in parts 3 and 4 to write a short memo for Ralf's boss explaining who called and what the call was about, and giving other key information.
6. How does Ralf handle the phone call? Discuss what he did well and what he could improve.

V 3 Job interview

lingua tv
we watch languages!

Before watching

1. Describe the photo.

While watching

2. Write down three of the questions that the interviewer asks.
3. Watch the film again and answer the questions.
   a. What is the applicant's name?
   b. What position is he applying for?
   c. What position did the applicant hold in his last job?
   d. According to the applicant, what is his greatest strength?
   e. What is the applicant's weakness?
   f. What questions does the applicant ask the interviewer?

After watching

4. Do you think the applicant will get the job? Give reasons for your answer.

V 4 Technical support

lingua tv
we watch languages!

While watching

1. What does 'rebooting the system' mean?
2. Watch the video again and decide whether the following statements are true or false.
   a. The caller cannot connect to the Internet.
   b. He has been having problems for two days now.
   c. He has tried rebooting the system.
   d. Turning the modem on and off did not help the problem.
   e. The scanner is not connected to the caller's computer.
   f. The technical support employee is unable to solve the problem with the scanner.

After watching

3. Make a list of problems that employees may have with their computers.
4. Record the call by copying the headings onto a sheet of paper and filling in the information.

Name of caller • Problem(s) • Solution(s) • Further action necessary: Yes/No

V 5 Waste – the future's most valuable resource

Before watching

1. What things do you recycle every day, and how?
2. How many different waste bins does your city or community provide per household?

While watching

3. Watch the film and find out the following information:
   a. How many tons of rubbish does Germany produce each year?
   b. What nations are eager to pay for our rubbish?
   c. What metal is particularly desirable for recycling?

4. Watch the film again and answer the following questions:
   a. How many tons of copper were stolen in Germany?
   b. How much heat can one ton of residual waste produce?

After watching

5. Explain these phrases/sentences from the film in your own words.
   a. "modern urban warfare"
   b. "Our waste is getting a second lease on life."

V 6 Pipe inspection

Before watching

1. What do pipelines transport?

While watching

2. Watch the film and find out the following information:
   a. What is the inspection device called?
   b. How long is it?
   c. How fast does the device travel?
   d. How many days does it take to inspect 300 km of pipeline?

3. Watch the film again and answer the following questions:
   a. Where are the data analyzed?
   b. How accurate is the data grid?
   c. What does the device use to identify defects?

After watching

4. What may cause defects in or damage to pipelines?

V7 High voltage work

Before watching

1. What do you think the men in the steel basket are preparing to do?

While watching

2. Watch the film and find out the following:
   a. what weather conditions are necessary for repairing overhead power lines.
   b. how many volts of electricity high-voltage lines carry.
   c. how the repairmen get to the damaged lines.
   d. what safety measures are taken.

3. Watch the film again and answer the following questions:
   a. How long does the job shown in the film take?
   b. Where is the electricity switched off and on?

After watching

4. Why is it important that the men work quickly?
5. Explain in your own words: "A power cut can quickly bring an entire city to its knees."

V8 Ras Laffan

While watching

1. Watch the film and find out the following:
   a. where Ras Laffan is located.
   b. what is processed there.
   c. how much is processed each year.
   d. how large the Ras Laffan industrial region is.

2. Watch the film again and answer the following:
   a. What problems do the construction workers face?
   b. How is the concrete cooled, and why?
   c. Why is the gas cooled?

After watching

3. Explain the phrase "liquid gold".

V9 Pipelines

Before watching

1. Name some things which are transported in pipelines.

While watching

2. Watch the film and find out the following:
   a. what alloy is popular for making long-distance pipelines.
   b. how thick are the steel plates that are formed into round pipes.
   c. the welding technique used in the factory shown.
   d. what the expander is for.

3. Watch the film again and answer the following:
   a. What conditions do offshore pipelines have to cope with?
   b. Why do they have an additional outer casing of concrete?

After watching

4. Explain why quality-control checks are important in the production of pipelines.

V10 Robots in the hospital

While watching

1. Watch the film and find out the following:
   a. What hospital is shown as an example? Which parts of the hospital do we see?
   b. How many robots do they have?
   c. What powers the robot couriers and how do they navigate?
   d. What is a 'lift-and-tunnel system'?

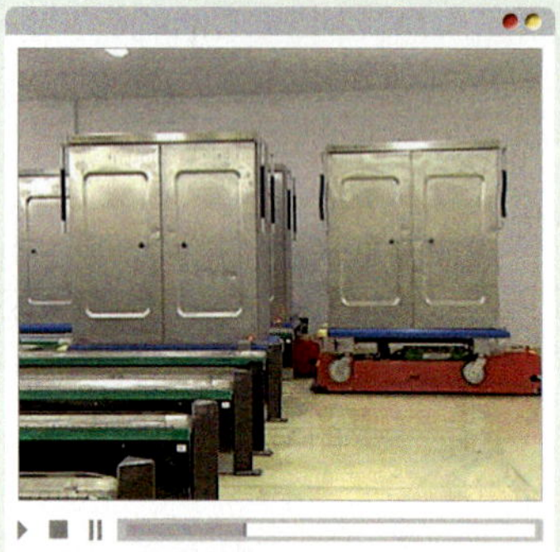

2. Watch the film again and answer these questions:
   a. What metal in the plates keeps the food warm?
   b. Do the robots distribute the food to the patients?
   c. How many containers are used to transport food?
   d. How does the hospital director feel about using robots to transport patients?

After watching

3. What else do you think the robots transport in the hospital?
4. Where else might transportation robots be used?

V11 Portable power stations

BBC Motion Gallery

Before watching

1. Think of situations in which a portable power station like the one in the photo could be useful.

While watching

2. Watch the film and list the uses of electrical power 'out in the open' that the reporter mentions.

3. Watch the film again and answer the following:
   a. What supplies the power for the 12-volt portable power station?
   b. Why are batteries not such a good choice?
   c. Where was the portable power station invented?

After watching

4. Do you think the inventor will be able to sell his power station in Germany? Where might the need for it be greater?

V12 Car speed control

BBC Motion Gallery

Before watching

1. What speed limit is acceptable
   a. on city streets?
   b. on the motorway?

While watching

2. Watch the film and find out the following:
   a. how many people die in road accidents in the UK every year.
   b. what the MIRA is working on.
   c. what the satellite positioning system does.

3. Watch the film again and answer the following:
   a. Is speed control popular with car manufacturers? Why or why not?
   b. Do speed cameras have any effect on driving habits?
   c. How is the speed-control system being tested?

After watching

4. Would you buy a car with a speed-control system? Give reasons for your answer.

KMK-Prüfungssätze

Stufe I *(100 VP)*

1 Hörverstehen **20 VP**

◎ A 6.1 **Sie interessieren sich für eine Stelle im englischsprachigen Ausland, z. B. ein Praktikum oder eine längerfristige Anstellung nach der Gesellen- oder Facharbeiterprüfung. Die Virtual Employment Agency (VEA) vermittelt solche Praktika. Sie laden einen Podcast mit vier Stellenangeboten herunter. Erstellen Sie hierzu eine Tabelle mit folgenden Stichwörtern und hören Sie den Podcast zweimal an. Füllen Sie Ihre Tabelle auf Englisch aus.**

Information number	1 *(5 VP)*	2 *(5 VP)*	3 *(5 VP)*	4 *(5 VP)*
trade/position offered				
length of practical				
company and location of practical				
requirements				
extra information: email address, telephone number if given				

2 Rezeption **20 VP**

Sie möchten Ihren Arbeitgeber überzeugen, Sie während Ihrer Ausbildung für ein Praktikum im Ausland freizustellen und haben dazu folgendes Informationsmaterial von der Internetseite der VEA heruntergeladen. Ihr Arbeitgeber bittet Sie, folgende Fragen zu diesem Text auf Deutsch zu beantworten.

1. Was erfahren wir über die Organisation der VEA? *(2 VP)*
2. Welche Vorteile eines Praktikums werden genannt? *(3 VP)*
3. Was konkret bietet die VEA an? *(3 VP)*
4. Was wird über eine Bezahlung/Kostenzuschuss für die Praktikanten gesagt? *(3 VP)*
5. Wie lange dauern die angebotenen Praktika? *(3 VP)*
6. Wer kann sich bei der VEA bewerben und welche Voraussetzungen sollten die Bewerber mitbringen? *(3 VP)*
7. Wie bewirbt man sich? Welche Unterlagen müssen geschickt werden? *(3 VP)*

London-based VEA offers job placements

Who we are

VEA is an organisation funded by the European Union and European companies. Our aim is to help to make European companies fit to operate more effectively in a global world by giving their trainees a chance to work abroad, improve their English language skills and become acquainted with a new culture. The trainees who take part in the programme have a chance to see different work techniques and learn to adapt to new working conditions.

▶

What we offer
VEA helps trainees to find work placements in England, Scotland and Wales. We have lists of companies in all kinds of trades who offer work placements. All placements take place in an authentic work environment. We can help you to establish contacts with companies and to plan your stay. We have a telephone hotline if problems should come up.
What payment is given
VEA does not pay trainees any money for the work placement, but we can pay a fixed amount towards your travelling expenses. However, most of the companies that we have under contract offer small salaries that will enable you to cover daily expenses such as food and board.
How long do work placements last?
The duration of work placements depends on the companies. Placements usually last between eight and ten weeks, but most companies are flexible if you want to arrange different periods. The shortest possible placement is three weeks.
Who can apply
All trainees can apply after their first or second year of training. The applicants should have a fair working knowledge of English. They should be communicative, interested in our culture and willing to participate in social activities if these are offered by the company.
How to apply
Send an email and a CV as an attachment in English telling us why you would like to take part in our work placement programme. You should also send written consent and a letter of recommendation from your company. And don't forget to mention the dates you would like to do the placement and the trade you would like to work in. Send your application by email to info@vea…biz subject matter "work placement."

(363 words)

3 Produktion **30 VP**

Sie wollen sich per E-Mail bei der VEA um ein Praktikum bewerben (Adresse: info@vea…biz). Schreiben Sie, dass Ihre Firma es lieber hätte, wenn Sie zunächst einmal nur 3 Wochen auf ein Auslandspraktikum gingen. Geben Sie an, welchen Beruf Sie lernen und in welchem Jahr Ihrer Ausbildung Sie sich befinden. Nennen Sie einen Zeitraum von drei Wochen (mit genauem Datum), in dem Sie ein Praktikum machen möchten.

4 Mediation **30 VP**

Auf der Internetseite befinden sich auch Erfahrungsberichte früherer Absolventen. Einer hat Ihnen besonders gefallen. Sie übertragen ihn ins Deutsche, um ihn ebenfalls Ihrem Arbeitgeber vorzulegen. Auf eine wörtliche Übersetzung legt er keinen Wert, lediglich auf eine inhaltlich korrekte Wiedergabe.

My name is Michael. I'm in my third year of training as a car mechanic. I've just come back from a three-month placement in South Wales. It was sponsored by the Leonardo Programme. My company in Germany ▶

encouraged me to apply for the placement. They even sponsored me by paying my wages even though I wasn't working for them for three months.

I worked in a company based near Cardiff that makes parts for the automobile industry. First I spent a few days in each department of the company, and then I was fully involved in the production process.

I worked with a Welsh colleague in the CNC centre. Of course I had some experience with CNC from my own company. I must say, I learned a lot, especially how to work more efficiently and independently.

In the beginning the language was a bit of a problem, but after a couple of weeks we managed to communicate quite well, and I learned many technical words and phrases.

This placement was a great opportunity for me, and my boss sees it as a real plus for our company and our business with British customers.

(193 words)

KMK Stufe II *(100 VP)*

1 Hörverstehen **8 VP**

⦾ A 6.2 **Ihre Firma stellt Hydraulikaggregate her. Sie werden gebeten, eine Nachricht auf dem Anrufbeantworter Ihrer Firma abzuhören und eine kurze Telefonnotiz auf Deutsch zu erstellen.**

Bevor Sie die Nachricht abhören, bereiten Sie auf einem Blatt eine Telefonnotiz mit folgenden Überschriften vor.

1. Name des Anrufers und der Firma *(2 VP)*
2. Sitz der Firma *(1 VP)*
3. Nachricht für … *(1 VP)*
4. Grund des Anrufes *(2 VP)*
5. Auftragsnummer *(1 VP)*
6. Rückrufnummer *(1 VP)*

2 Hörverstehen **12 VP**

⦾ A 6.3 **Für die Herstellung von Steuerungen und Motoren sind seltene Erden unverzichtbar. Ihr Ausbildungsleiter hat zu seltenen Erden einen englischen Podcast gefunden. Da der Ausbilder nicht so gut Englisch kann, bittet er Sie, sich den Podcast anzuhören und folgende Fragen auf Deutsch zu beantworten.**

1. Was wird über den Bedarf an seltenen Erden in der Einführung gesagt? *(1 VP)*
2. Welche Hinweise werden in der Einführung über die Rolle Chinas gegeben? *(2 VP)*
3. Ihr Ausbildungsleiter hat nicht genau verstanden, ob und wie seltene Erden wiederverwertet werden können. Notieren Sie die Aussage des Sprechers auf Deutsch. *(1 VP)*
4. Der Podcast nennt einige Bereiche, in denen seltene Erden unverzichtbar sind. Nennen Sie vier dieser Bereiche. *(2 VP)*
5. In welchem Bereich werden gleich mehrere Kilogramm an seltenen Erden pro Einheit gebraucht? *(1 VP)*

6. Wie selten sind seltene Erden nach Auskunft des Podcasts wirklich? *(1 VP)*
7. Es werden Produktionszahlen für China genannt: Wie hoch ist der Anteil der Chinesen an den geschätzten Reserven weltweit? *(1 VP)*
8. Welche anderen Länder werden genannt, in denen seltene Erden gefördert werden? *(2 VP)*
9. Welche Sorgen plagen das amerikanische Repräsentantenhaus? *(1 VP)*

3 Rezeption

20 VP

Im Umgang mit Druckluft müssen bestimmte Sicherheitsregeln beherzigt werden. Bei einem Druckluftwerkzeug, das Ihre Firma geliefert bekam, war folgendes Faltblatt beigefügt. Ihr Ausbilder stellt Ihnen dazu ein paar Fragen, die Sie anhand dieses Textes auf Deutsch auf einem gesonderten Blatt beantworten sollen.

1. Der Text zählt eine Reihe von Druckluftwerkzeugen auf. Nennen Sie acht davon in der Reihenfolge, in der der Text sie aufführt. *(4 VP)*
2. Welche Sicherheitsausrüstung muss laut Text benutzt werden? *(2 VP)*
3. Was muss zum Schutz Ihrer Mitarbeiter getan werden? *(2 VP)*
4. Wie sind laut Text Druckluftwerkzeuge zu behandeln? *(3 VP)*
5. Was wird über die Zusatzgeräte ausgesagt? *(1 VP)*
6. Welche allgemeinen Eigenschaften sollen die Schläuche haben (bitte in der Reihenfolge aufführen wie im Text)? *(2 VP)*
7. Welche speziellen Eigenschaften hinsichtlich des Druckes sind bei den Schläuchen einzuhalten? *(2 VP)*
8. Was müssen Sie laut Text tun, bevor Sie Werkzeuge an den Druckluftschlauch anschließen? *(1 VP)*
9. Welchen Anforderungen müssen die Schlauchverbindungen entsprechen? *(2 VP)*
10. Druckluft wird gerne auch zur Reinigung verwendet. Was wird über die Reinigung mit Druckluft in diesem Text ausgesagt? *(1 VP)*

What are pneumatic tools?

- Pneumatic tools are powered by compressed air. Common types of these air-powered hand tools that are used in industry include buffers, nailing and stapling guns, grinders, drills, jack hammers, chipping hammers, riveting guns, sanders and wrenches.

How do you use pneumatic tools safely?

- Review the manufacturer's instruction before using a tool.
- Wear safety glasses or goggles, or a face shield and, where necessary, safety shoes or boots and hearing protection.
- Post warning signs where pneumatic tools are used. Set up screens or shields in areas where nearby workers may be exposed to flying fragments, chips, dust, and excessive noise.
- Ensure that the compressed air supplied to the tool is clean and dry. Dust, moisture, and corrosive fumes can damage a tool. An in-line regulator filter and lubricator increases tool life.
- Keep tools clean and lubricated, and maintain them according to the manufacturer's instructions.

▶

- Use only the attachments that the manufacturer recommends for the tools you are using.
- Be careful to prevent hands, feet, or body from injury in case the machine slips or the tool breaks.
- Reduce physical fatigue by supporting heavy tools with a counter-balance wherever possible.

How should you handle air hoses?
- Use the proper hose and fittings of the correct diameter.
- Use hoses specifically designed to resist abrasion, cutting, crushing and failure from continuous flexing.
- Choose air-supply hoses that have a minimum working pressure rating of 1035 kPa (150 psig) or 150 % of the maximum pressure produced in the system, whichever is higher.
- Check hoses regularly for cuts, bulges and abrasions. Tag and replace, if defective.
- Blow out the air line before connecting a tool. Hold hose firmly and blow away from yourself and others.
- Make sure that hose connections fit properly and are equipped with a mechanical means of securing the connection (e.g. chain, wire, or positive locking device).
- Install quick disconnects of a pressure-release type rather than a disengagement type. Attach the male end of the connector to the tool, NOT the hose.
- Do not operate the tool at a pressure above the manufacturer's rating.
- Turn off the air pressure to hose when not in use or when changing power tools.
- Do not carry a pneumatic tool by its hose.
- Avoid creating trip hazards caused by hoses laid across walkways or curled underfoot.
- Do not use compressed air to blow debris or to clean dirt from clothes.

What should you avoid with compressed air?
- Cleaning with compressed air is dangerous.
- Do not use compressed air for cleaning unless no alternate method of cleaning is available. The nozzle pressure MUST remain below 207 kPa (30 psi). Personal protective equipment and effective chip guarding techniques must be used. (...)

(489 words)

4 Mediation

30 VP

Im Lernfeldunterricht Ihrer Berufsschule bekommen Sie folgenden (siehe nächste Seite) englischen Text vorgelegt. Lesen Sie ihn durch und übertragen Sie die wichtigsten Punkte in zusammenhängenden Sätzen ins Deutsche. Eine wörtliche Übersetzung ist nicht gewünscht.

Processes for joining metals

When metals are to be joined different processes can be applied such as welding, soldering and brazing. There are two kinds of welding: gas welding and electric-arc welding. By definition, welding is a fusion process that joins materials of the same kind, usually metals or thermoplastics. This is often done by melting the workpieces and adding a filler to form a pool of molten material that cools to become a strong joint; pressure is sometimes used together with high temperature, up to 3000°C, to produce the weld.

This is in contrast to soldering and brazing, which use a material with a lower melting point between the workpieces to form a bond between them without melting the workpieces.

The solder used in soldering has a melting point of up to 450°C. The parts which are to be joined are heated, and the solder is pressed against them. The solder melts and flows into the joint. After the material has cooled down, the joint is finished.

Higher temperatures are needed in brazing: the solder has a melting point of between 450°C and 900°C. Metals suitable for brazing are steel, copper, silver and brass.

Electrical engineers apply soldering to join electronic parts onto a PCB (printed circuit board), for example. Plumbers use this joining technique to fit water or gas pipes, to join drain pipes or gutters, or to generate roof sealings.

(233 words)

5 Produktion **30 VP**

Ihre Firma plant einen Messeauftritt auf der *Automotive Testing Expo* in Shanghai. Zur Vorbereitung hatte sie zwei Monate vor dem Messetermin den Besuch eines Mitarbeiters in Shanghai geplant, um mit chinesischen Zulieferern den Auftritt zu besprechen. Leider wurde der Termin von chinesischer Seite kurzfristig um zwei Wochen verschoben. Schreiben Sie eine E-Mail mit folgenden Punkten an das Royal Victoria Hotel in Shanghai info@royalvictoria. . .com .

- Bedanken Sie sich für die Reservierung eines Einzelzimmers mit Dusche, Satellitenfernseher, schnurlosem Telefon und Internetanschluss für Nichtraucher sowie für die Reservierung eines Sitzungsraumes für kurze Geschäftsbesprechungen für den Zeitraum 01.–08.08.20…
- Teilen Sie mit, dass Sie die Reservierungen leider stornieren müssen, weil Ihre Geschäftsreise um zwei Wochen verschoben werden musste.
- Fragen Sie an, ob Sie Ihre Buchungen ebenfalls um zwei Wochen verschieben können, und zwar auf den 15.–22.08.… .
- Bitten Sie um eine schriftliche Bestätigung der neuen Buchungen.
- Bitten Sie um die Empfehlung eines anderen Hotels für den Fall, dass das Royal Victoria Hotel zu dem gewünschten Termin keine Zimmer mehr frei hat.
- Denken Sie beim Verfassen Ihrer E-Mail auch an die äußere Form der E-Mail wie Betreff, Anrede, Schluss und Grußformel.

Role cards

Module 2 Tools in the car workshop, A 3

Partner A (questions)

Use the following information to ask your partner questions. You start.

1. Was würdest du benutzen, um Schrauben im Innenraum eines Fahrzeugs anzuziehen?
2. OK, danke. Ich will zudem den Sitz ausbauen. Er ist mit Sechskanntschrauben fixiert. Welches Werkzeug würdest du empfehlen?
3. Ja, in Ordnung. Noch eine letzte Frage. Ich muss ein Metallstück kürzen. Welche Werkzeuge sollte ich dafür verwenden?

Module C Telephoning, B 3

→ **PHRASES:** Telephoning

Partner A

Give your partner the following information:

1. The room is 54 m^2.
2. The wire needs to be 128.5 cm long.
3. The doors are 80 cm x 220 cm.
4. Each of the components weighs around 0.75 kg.

Module 10 Wheels and brakes, A 5

→ **PHRASES:** Dealing with customers

Partner A: Sales assistant

You are a sales assistant who specialises in selling car tyres. Use the information below to role-play a sales dialogue with a customer. You start.

- Welcome the customer and find out how you can help.
- Ask for vehicle documents.
- Tell the customer that he / she needs tyres size 235 / 45 R17 91H.
- Explain that '235' stands for the tyre width in mm and the letter 'H' for permissible top speed (in this case 210 km / h).
- Price for a set of four tyres is 550 €.
- Agree to a maximum 10% discount *(Rabatt)* but the customer must pay in cash.
- Say that you can fit the tyres next Monday. If the customer does not accept, see whether you can get him / her to agree if you fit the tyres immediately.

Module 2 Tools in the car workshop, A 3

Partner B (answers)

Use the following information to answer your partner's questions. Partner A starts.

1. Ich würde einen Kreuzschlitzschraubenzieher benutzen. Falls es eine Schlitzschraube ist, dann nimm einen Schlitzschraubenzieher.
2. Es gibt verschiedene Möglichkeiten. Du könntest z. B. einen Schraubenschlüssel benutzen. Einfacher und schneller geht es aber mit einen Steckschlüssel und einer Ratsche.
3. Ich würde das Metallstück mit einer Gripzange fixieren und dann kannst du mit einer Metallsäge das Stück absägen.

Module C Telephoning, B 3

→ **PHRASES:** Telephoning

Partner B

Give your partner the following information:

1. The table is 2 m x 3 m.
2. The engine runs at 7500 rpm.
3. 220 V sockets are standard in Germany.
4. The customer would like the work surface to be 24 mm thick.

Module 10 Wheels and brakes, A 5

→ **PHRASES:** Dealing with customers

Partner B: Customer

You are a customer and you need new tyres for your car. Use the information below to role-play a dialogue with a sales assistant. Partner A starts.

- Introduce yourself and tell the sales assistant you want to buy a set of new summer tyres.
- Give the car documents to the sales assistant.
- Write down the size of the tyres and ask what the first number '235' and the last letter 'H' stand for.
- Find out the price.
- You think the tyres are too expensive and ask for a 15% discount *(Rabatt)*. You want to pay in cash.
- Decide whether you can agree to the assistant's offer and ask when the tyres could be fit if you decide to buy them.

Phrases

A | Meeting people

Formal greetings	Formelle Begrüßung
Good morning / afternoon / evening, Ms Robinson / Mr Jones.	Guten Morgen / Tag / Abend, Frau Robinson / Herr Jones.
Good morning / afternoon / evening, ladies and gentlemen.	Guten Morgen / Tag / Abend, meine Damen und Herren.
Informal greetings	**Informelle Begrüßung**
Good morning, Susan / James.	Guten Morgen, Susan / James.
Hi / Hello / Morning, Karl.	Tag / Morgen, Karl.
Hi everybody!	Tag zusammen!
Formal goodbyes	**Formelle Verabschiedung**
Goodbye! It's been nice / pleasant meeting you.	Auf Wiedersehen! Es war nett / angenehm, Sie kennen zu lernen!
Thank you for your help / your time. Let's keep in touch.	Danke für Ihre Hilfe / Ihre Zeit. Wir bleiben in Kontakt.
Informal goodbyes	**Informelle Verabschiedung**
Bye!	Tschüss / Ciao!
See you later!	Bis später!
See you around!	Wir sehen uns!
See you tomorrow / on Monday / next week!	Bis morgen / Montag / nächste Woche!
Formal introduction	**Formelle Vorstellung**
May I introduce myself?	Darf ich mich vorstellen?
Hello, my name is / I'm Lukas Hansen from ATW Shipping.	Hallo, mein Name ist / ich heiße Lukas Hansen von ATW Shipping.
Pleased to meet you. / Nice to meet you. / How do you do?	Angenehm. / Sehr erfreut.
It's a pleasure to finally meet you in person.	Es freut mich sehr, Sie endlich einmal persönlich kennen zu lernen.
Informal introduction	**Informelle Vorstellung**
Hi. I'm Peter, Peter James.	Hi. Ich bin Peter, Peter James.
Hello, my name is Sarah O'Neill.	Hallo, ich heiße Sarah O'Neill.
Hi Sarah. Nice to meet you.	Hi Sarah. Freut mich, dich kennen zu lernen.

Introducing others (formal)	Andere miteinander bekannt machen (formell)
Have you met our managing director?	Haben Sie schon unsere Geschäftsführerin / unseren Geschäftsführer kennen gelernt?
Ms Hall? May I introduce you to Christian Eriksen, our new intern? Christian, Georgina Hall from our sales department.	Frau Hall? Darf ich Sie Christian Eriksen, unserem neuen Praktikanten, vorstellen? Christian, das ist Georgina Hall von unserer Verkaufsabteilung.
Introducing others (informal)	**Andere miteinander bekannt machen (informell)**
Here's someone I'd like you to meet.	Ich möchte euch gerne jemanden vorstellen.
Christian, this is Joe. Joe, Christian.	Christian, das ist Joe. Joe, das ist Christian.
Let me introduce you to the team.	Darf ich dich dem Team vorstellen?
Use of first names	**Verwendung von Vornamen**
Please call me Martin.	Bitte nennen Sie mich ruhig Martin.
My name's Marianne.	Ich heiße Marianne.
My first name's Andrew, but everybody calls me Andy.	Mein Vorname ist Andrew, aber alle nennen mich Andy.
May I call you Anna?	Darf ich Anna zu Ihnen sagen?

B | Your company

Introducing the company	Die Firma vorstellen
Our company was founded in 1997.	Unsere Firma wurde 1997 gegründet.
In 2005 we were taken over by …	2005 wurden wir von … übernommen.
We are a leading manufacturer of engines.	Wir sind ein führender Hersteller von Motoren.
We are a medium-sized family firm.	Wir sind ein mittelständisches Familienunternehmen.
We are a manufacturer of industrial cleaning machines specializing in high-pressure cleaners.	Wir sind ein Hersteller von industriellen Reinigungsgeräten und sind auf Hochdruckreiniger spezialisiert.
I work for a major construction company.	Ich arbeite bei einem bedeutenden Bauunternehmen.
We are a subsidiary of …	Wir sind eine Tochtergesellschaft von …
We are a chain of electronics suppliers.	Wir sind eine Kette von Elektronik-Anbietern.
Describing the company's products / services	**Die Produkte / Dienstleistungen der Firma beschreiben**
We offer sustainable solutions.	Wir bieten nachhaltige Lösungen an.
Our products are environmentally friendly.	Unsere Produkte sind umweltfreundlich.
Our products are well-known in Europe.	Unsere Produkte sind europaweit bekannt.
Our components are both reliable and long-lasting.	Unsere Komponenten sind zuverlässig und haben eine lange Lebensdauer.
Our software is adapted to suit your requirements.	Unsere Software wird Ihren Bedürfnissen angepasst.
We manufacture high-tech shock absorbers.	Wir stellen High-Tech-Stoßdämpfer her.
Our products are designed using the latest CAD technology.	Unsere Produkte werden mit der neuesten CAD-Technologie entworfen.

Describing the job	Den Beruf beschreiben
I'm training to become a ...	Ich mache eine Lehre / Ausbildung zur / zum ...
I am a bricklayer / carpenter / technician / plumber / joiner.	Ich bin Maurer(in), Zimmermann / Zimmerin, Techniker(in), Klempner(in), Schreiner(in).
I work in the automotive industry.	Ich arbeite in der Automobilindustrie.
I work with automated machines.	Ich arbeite mit automatisierten Maschinen.
Describing the workplace	**Den Arbeitsplatz beschreiben**
I work on an assembly line / a machine.	Ich arbeite am Fließband / an einer Maschine.
I work in a large warehouse / factory.	Ich arbeite in einem großen Warenlager / einer großen Fabrik.
I work in a garage / workshop.	Ich arbeite in einer Autowerkstatt / Werkstatt.
I work on a building site.	Ich arbeite auf einer Baustelle.
Talking about tasks and responsibilities	**Über Aufgaben und Zuständigkeiten sprechen**
I am responsible for programming the machines.	Ich bin für die Programmierung der Maschinen zuständig.
I report to the foreman.	Ich bin dem Meister / Vorarbeiter unterstellt.
I work shifts.	Ich mache Schichtarbeit.
I often have to do overtime.	Ich muss oft Überstunden machen.
I repair electrical devices.	Ich repariere elektrische Geräte.

C | Telephoning

1. Answering the phone	Sich am Telefon melden
Introducing yourself	**Sich vorstellen**
Smart EDV GmbH, good morning. Hendrik Klopp speaking.	Smart EDV GmbH, guten Morgen. Hendrik Klopp am Apparat.
Offering help	**Hilfe anbieten**
How may I help you? / Can I help you?	(Wie) kann ich Ihnen helfen?
What can I do for you?	Was kann ich für Sie tun?
Asking for details	**Nach Details fragen**
Who's calling, please?	Wer ist am Apparat, bitte?
May I / Can I ask who's calling, please?	Darf ich fragen, wer anruft?
May I / can I take your name, please?	Könnten Sie mir bitte Ihren Namen nennen?
Could you spell that please?	Könnten Sie das bitte buchstabieren?

Connecting	Verbinden
One moment, please. I'm putting you through now.	Einen Moment bitte, ich verbinde Sie.
Shall I put you through to …?	Soll ich Sie mit … verbinden?
Hold the line, please. I'll put you through.	Bitte bleiben Sie am Apparat. Ich verbinde.
Your boss is not available	**Der / die Vorgesetzte ist nicht zu sprechen**
I'm afraid Mr Deschler is not in the office at the moment.	Es tut mir leid, Herr Deschler ist z. Zt. leider nicht in seinem Büro.
… He is on a building site.	… Er ist auf einer Baustelle.
… He has a visitor with him.	… Er hat Besuch.
… He is out at lunch.	… Er ist in der Mittagspause.
Offering to call back	**Einen Rückruf anbieten**
Shall I ask her to call you back?	Soll ich sie bitten, Sie zurückzurufen?
Can she call you back this afternoon?	Kann sie Sie heute Nachmittag zurückrufen?
Does he have your telephone number?	Hat er Ihre Telefonnummer?
Can I just confirm your number?	Kann ich Ihre Nummer nochmals überprüfen?
I'll just repeat that.	Ich wiederhole.
Taking a message	**Eine Nachricht aufnehmen**
Can I take a message?	Kann ich etwas ausrichten?
Can I give him a message?	Kann ich ihm etwas ausrichten?
Would you like to leave a message?	Möchten Sie eine Nachricht hinterlassen?
Can I just confirm that?	Kann ich das kurz überprüfen?
So that's twenty-four square metres, isn't it?	Das sind also vierundzwanzig Quadratmeter, oder?
I'll make sure he gets the message.	Ich werde mich darum kümmern, dass er die Nachricht bekommt.
I'll tell her as soon as she comes back.	Ich sage ihr Bescheid, sobald sie zurückkommt.
Problems	**Probleme**
Sorry, I didn't quite catch that. Could you repeat it more slowly, please?	Es tut mir leid, das habe ich nicht verstanden. Könnten Sie es bitte etwas langsamer wiederholen?
I'm afraid I didn't catch the telephone number. Could you repeat it, please?	Leider habe ich die Telefonnummer nicht mitbekommen. Würden Sie sie bitte wiederholen?
I'm sorry, but we are having a bad connection. Could you repeat that, please?	Es tut mir leid, aber die Verbindung ist schlecht. Könnten Sie das bitte nochmals wiederholen?
Ending the call	**Das Telefongespräch beenden**
You're welcome.	Bitteschön. / Gern geschehen.
Thanks for calling.	Vielen Dank für Ihren Anruf.
Goodbye.	Auf Wiederhören.

2. Making a call	Jemanden anrufen
Introducing yourself	**Sich vorstellen**
This is … from Fertighaus Henke.	Hier spricht … von Fertighaus Henke.
My name's … I'm calling from Smart EDV.	Mein Name ist … . Ich rufe von der Firma Smart EDV an.
Asking to speak to somebody	**Nach jemandem fragen**
Could I speak to Ms Sheringham, please?	Könnte ich bitte Frau Sheringham sprechen?
Could you put me through to Mr Ferguson?	Könnten Sie mich mit Herrn Ferguson verbinden?
Could you give me his / her extension number, please?	Könnten Sie mir bitte seine / ihre Durchwahl geben?
I'd like to speak to someone from the production planning department.	Ich möchte mit jemandem in der Fertigungsplanung sprechen.
Spelling your name	**Den Namen buchstabieren**
It's S-C-H-N-E-I-D-E-R. Schneider.	Das ist S-C-H-N-E-I-D-E-R. Schneider.
Giving your phone number	**Die Telefonnummer nennen**
It's 0049 for Germany, then 0391 458 547.	Das ist die 0049 für Deutschland, dann 0391 458 547.
It's 0391 458 547.	Das ist die 0391 458 547.
Leaving a message	**Eine Nachricht hinterlassen**
Yes, please. Could you tell him that …?	Ja bitte. Könnten Sie ihm ausrichten, dass …?
Thank you. I'll ring back later.	Danke sehr. Ich rufe später zurück.
I'm afraid I won't be in the workshop this afternoon. I'll give you my mobile number.	Leider bin ich heute Nachmittag nicht in der Werkstatt. Ich gebe Ihnen meine Handynummer.
Ending the call	**Das Telefongespräch beenden**
Thank you very much for your help.	Vielen Dank für Ihre Hilfe.
Goodbye.	Auf Wiederhören.

D | Written communication

Salutation	Anrede
Dear Sir or Madam	Sehr geehrte Damen und Herren
Dear Mr … / Dear Ms …	Lieber Herr … / Liebe Frau …
Dear Peter	Lieber Peter
Complimentary close	**Schlussformel**
Yours sincerely / Yours faithfully (rarely used)	Mit freundlichen Grüßen / Hochachtungsvoll (selten benutzt)
Best regards / Kind regards / Best wishes	Schöne Grüße / Herzliche Grüße

Enquiries	Anfragen
We have visited your website and ...	Wir haben Ihre Webseite besucht und ...
Your services have been recommended to us by ...	Ihre Dienstleistungen wurden uns von ... empfohlen.
We are a well-established manufacturer of ...	Wir sind ein gut etablierter Hersteller von ...
Our firm is a leading importer of ...	Unsere Firma ist ein führender Importeur von ...
Could you please let us have a brochure and a price list?	Wir bitten um einen Prospekt und eine Preisliste.
Please send us a quotation for ...	Bitte machen Sie uns ein Angebot über ...
Please quote your lowest prices for ...	Bitte nennen Sie uns Ihre günstigsten Preise für ...
We would be grateful for information on your terms of payment and delivery.	Wir bitten um nähere Angaben zu Ihren Liefer- und Zahlungsbedingungen.
A visit by your representative would be appreciated.	Wir wären dankbar für einen Besuch Ihres Vertreters.
We look forward to hearing from you.	Wir freuen uns darauf, bald von Ihnen zu hören.
Offers	**Angebote**
Many thanks for your enquiry of 30 May about our new range of ...	Wir danken Ihnen vielmals für Ihre Anfrage vom 30. Mai wegen unseres neuen Sortiments von ...
As requested, we are sending you enclosed our latest catalogue and price list.	Wie gewünscht fügen wir unseren neuesten Katalog und unsere Preisliste bei.
We can offer a 10% quantity discount on orders of at least 500 units.	Für Aufträge über mindestens 500 Stück wird 10% Mengenrabatt gewährt.
We take pleasure in submitting the following cost estimate:	Wir freuen uns, Ihnen folgenden Kostenvoranschlag zu unterbreiten:
Our usual terms of payment are: – cash with order – cash on delivery – 30 days net, 10 days 2% – by irrevocable letter of credit	Unsere üblichen Zahlungsbedingungen lauten: – Barzahlung bei Auftragserteilung – Barzahlung bei Lieferung – 30 Tage netto, 10 Tage 2% Skonto – durch unwiderrufliches und bestätigtes Akkreditiv
The delivery period is 6 weeks.	Die Lieferzeit beträgt 6 Wochen.
We look forward to welcoming you as our customers.	Wir freuen uns darauf, Sie als Kunden begrüßen zu dürfen.
Orders	**Bestellungen**
Please supply the following items on the terms stated below:	Bitte liefern Sie uns folgende Positionen zu den unten genannten Bedingungen:
As agreed, we will effect payment by bank transfer 30 days from date of invoice.	Wie vereinbart werden wir die Zahlung 30 Tage nach Rechnungsdatum per Banküberweisung vornehmen lassen.
Please note that the goods must reach us by 1 March at the latest.	Wir weisen darauf hin, dass die Ware bis spätestens 1. März hier eintreffen muss.
Please acknowledge this order promptly.	Bitte bestätigen Sie diesen Auftrag umgehend.
We look forward to receiving the goods in time and to doing further business with you.	Wir freuen uns, die Ware rechtzeitig entgegennehmen und weitere Geschäfte mit Ihnen machen zu können.

E | Applications

Mentioning the source of the advertisement	Die Quelle der Anzeige nennen
I would like to apply for the job / work placement which you advertised in the … of … / on the … website.	Ich möchte mich auf die in der … vom … / auf der … Website ausgeschriebenen Stelle / Praktikumsstelle bewerben.
I saw your advertisement for a … (job / work placement) in the … of … / on the … website and would like to apply for the position.	Ich habe Ihre Anzeige für eine Stelle / Praktikumsstelle in der … vom … / auf der … Website gesehen und möchte mich auf die Stelle bewerben.
I am writing in response to your advertisement of … in the … / on the … website for a … (job / work placement).	Ich melde mich auf Ihre Anzeige vom … in der … / auf der … Website für eine Stelle / Praktikumsstelle.
Introducing yourself	**Sich vorstellen**
I am particularly attracted to this position as …	Ich bin an dieser Stelle besonders interessiert, weil …
I am currently doing a one-year / two-year … apprenticeship at a vocational college in … (town), Germany.	Ich mache zurzeit eine ein-/zweijährige … Ausbildung an einer berufsbildenden Schule in …, Deutschland.
I am in the second year of an apprenticeship in …	Ich bin im zweiten Jahr einer Ausbildung als …
At present I am working as a … with … (company).	Zurzeit arbeite ich als … bei der Firma …
After completing my apprenticeship / At the end of the apprenticeship I will be a qualified (job title).	Nachdem ich meine Ausbildung beendet habe / Mit Abschluss meiner Ausbildung bin ich eine qualifizierte / ein qualifizierter (Berufsbezeichnung).
Talking about yourself and saying why you are a good applicant	**Über sich sprechen und begründen, warum man ein guter Bewerber / eine gute Bewerberin ist**
As you will see from my CV …	Wie aus meinem Lebenslauf zu ersehen ist, …
I have a good knowledge of …	Ich verfüge über gute Kenntnisse in …
I am motivated / reliable / hard-working / keen to learn / friendly / open / a good communicator.	Ich bin motiviert / zuverlässig / sehr fleißig / lernwillig / freundlich / aufgeschlossen / ein kommunikativer Mensch.
I have previous work experience as …	Ich verfüge bereits über Berufserfahrung als …
I have a wide range of interests including …	Ich bin vielseitig interessiert, u.a. an …
I am familiar with this kind of work, because …	Mit dieser Tätigkeit bin ich vertraut, da …
I would very much welcome the opportunity to work in …	Über die Möglichkeit, in … zu arbeiten, würde ich mich sehr freuen.
I believe that I would be a valuable / enthusiastic employee.	Ich bin überzeugt, dass ich ein(e) nützliche(r) / engagierte(r) Mitarbeiter(in) wäre.

Finishing the letter	Den Brief beenden
I hope that you will consider my application.	Ich hoffe, dass Sie meine Bewerbung in Betracht ziehen werden.
Thank you in advance for considering my application.	Ich danke Ihnen im Voraus, dass Sie meine Bewerbung in Betracht ziehen.
I look forward to hearing from you (soon).	Ich freue mich auf eine (baldige) Antwort.

F | Socialising

Ice breakers	Eisbrecher
How was your weekend? – Good, thanks. How about yours?	Wie war dein / Ihr Wochenende? – Gut, danke. Und bei dir / Ihnen?
Did you see the game last night? – Of course! It was unbelievable. What did you think about the red card / winning goal?	Hast du / Haben Sie das Spiel gestern abend gesehen? – Natürlich! Es war unglaublich. Was halten Sie von der roten Karte / vom Siegtor?
Quite cold / warm today, isn't it?	Ziemlich kalt / warm heute, oder?
I really like your watch. Where did you get it?	Ihre / deine Uhr gefällt mir wirklich gut. Woher haben Sie / hast du sie?
Did you read the story about … in the newspaper this morning?	Haben Sie heute Morgen den Bericht über … in der Zeitung gelesen?
Ah! I see you are also in need of a coffee / cup of tea.	Ah! Wie ich sehe brauchen Sie / brauchst du auch einen Kaffee / eine Tasse Tee.
Is this your first time in Germany?	Sind Sie / Bist du zum ersten Mal in Deutschland?
The food is pretty good here, isn't it? – You can say that again!	Das Essen hier ist ziemlich gut, oder? – Das können Sie / kannst du laut sagen.
Keeping the conversation going	**Das Gespräch in Gang halten**
Really? Is that right? You're joking!	Wirklich? Ist das wahr? Sie machen / du machst Scherze!
That's brilliant / hilarious / fascinating!	Das ist brilliant / sehr lustig / faszinierend!
And what happened then?	Und was passierte dann?
No, unfortunately I missed it. What happened?	Nein, leider habe ich es verpasst. Was ist passiert?
It's interesting you should say that, because …	Interessant, dass Sie das sagen, weil …
Talking about food, have you already tried the new Italian restaurant in the city centre?	Wo wir gerade über Essen sprechen: Haben Sie schon den neuen Italiener in der Innenstadt ausprobiert?
Changing the subject	**Das Thema wechseln**
I think it's best if we don't talk about politics.	Ich denke, es ist das Beste, wenn wir nicht über Politik sprechen.
If you don't mind, I'd rather not talk about that.	Wenn es Ihnen nichts ausmacht, würde ich lieber nicht darüber sprechen.
Yes … By the way, have you read …	Ja … Übrigens, haben Sie … gelesen?

Ending the conversation politely	Die Unterhaltung höflich beenden
Well, it's been good talking to you.	Nun, es war schön, mit Ihnen zu sprechen.
I suppose it's time to get back to work. See you around.	Ich denke, dass es Zeit ist, zur Arbeit zurückzukehren. Man sieht sich!
Excuse me, but I have just seen someone I need to speak to. It was nice talking to you.	Entschuldigen Sie, aber ich habe jemanden gesehen, mit dem ich sprechen muss. Es war nett, mit Ihnen zu reden.
Eating out	**Essen gehen**
Talking about preferences	**Über Vorlieben sprechen**
Do you eat meat / seafood / fish?	Essen Sie Fleisch / Meeresfrüchte / Fisch?
Do you like spicy food?	Mögen Sie scharfes Essen?
I'm a vegetarian / vegan.	Ich bin Vegetarier / Veganer.
I'm allergic to nuts.	Ich bin allergisch gegen Nüsse.
I'd prefer something light; I'm on a diet.	Ich würde etwas Leichtes bevorzugen – ich bin auf Diät.
I don't drink alcohol.	Ich trinke keinen Alkohol.
Talking about the menu	**Über die Speisekarte sprechen**
It's sweet / savoury / sour.	Es ist süß / herzhaft / sauer.
It's a starter / main dish / dessert.	Es ist ein(e) Vorspeise / Hauptgericht / Dessert.
What's in it?	Was ist da drin?
It's made of …	Es wird aus … gemacht.
It's boiled / fried / baked / roasted / filled with …	Es ist gekocht / gebraten/gebacken / geröstet / gefüllt mit …
Ordering	**Bestellen**
Are you ready to order?	Sind Sie bereit zum Bestellen?
As a starter, I'll take the … and then …	Als Vorspeise nehme ich …, und dann …
I'd like a side order of rice, please.	Ich hätte gerne (eine Portion) Reis als Beilage.
Shall we share a bottle of wine?	Sollen wir uns eine Flasche Wein teilen?
How do you like your steak? – Rare / medium / well-done, please.	Wie möchten Sie Ihr Steak? – Englisch / medium / durch, bitte.
What kind of wine / beer do you have?	Welche Weine / Biersorten haben Sie?
Talking about the food	**Über das Essen sprechen**
Enjoy your meal.	Guten Appetit!
How's your meal?	Wie ist Ihr Essen?
What's your … like?	Wie ist Ihr …?
The meat is very tender / a little tough.	Das Fleisch ist sehr zart / etwas zäh.
It's not as sweet as I expected.	Es ist nicht so süß, wie ich gedacht habe.
Excuse me, but my soup is cold.	Entschuldigen Sie, aber meine Suppe ist kalt.

Asking for the bill	Nach der Rechnung fragen
Excuse me, could we have the bill, please?	Entschuldigen Sie, könnten wir bitte die Rechnung haben?
Do you accept credit cards?	Akzeptieren Sie Kreditkarten?
This one's on me!	Das geht auf mich!
Shall we split the bill 50-50?	Sollen wir die Rechnung 50-50 teilen?

G | Presentations

Beginning the presentation	Die Präsentation beginnen
I should like to start by telling you about my company.	Zunächst möchte ich Ihnen etwas über meine Firma sagen.
My presentation will deal with …	Meine Präsentation behandelt …
I intend to keep my presentation as brief as possible.	Ich möchte meine Präsentation so kurz wie möglich halten.
I would welcome any questions at the end of my presentation.	Ich wäre gern bereit, etwaige Fragen am Ende meiner Präsentation zu beantworten.
The handout summarises the main points and gives an overview of the relevant figures and statistics.	Das Handout enthält die Hauptpunkte und gibt einen Überblick über die entsprechenden Zahlen und Statistiken.
Structuring the main part of your presentation	**Den Hauptteil der Präsentation strukturieren**
Now my second point is …	Ich komme nun zu Punkt 2 …
Thirdly, let me give you some basic statistics.	Drittens darf ich Ihnen ein paar grundlegende Statistiken zeigen.
I should now like to move on to the next topic.	Ich möchte nun gern zum nächsten Thema übergehen.
An excellent example of this is …	Ein hervorragendes Beispiel dafür ist …
I should like to give you an example to illustrate this point.	Ich möchte diesen Punkt mit einem Beispiel erläutern.
Concluding your presentation	**Die Präsentation beenden**
To sum up we can say that …	Zusammenfassend kann man sagen, dass …
I should like to finish by saying / thanking the organisers / pointing out …	Ich möchte schließen mit der Bemerkung / dem Dank an die Organisatoren / dem Hinweis …
Are there any questions?	Gibt es Fragen?
Thank you for your attention.	Vielen Dank für Ihre Aufmerksamkeit.

H | Dealing with customers

Making a complaint	Eine Beschwerde / Reklamation vorbringen
We are writing with reference to our order no. …	Wir nehmen Bezug auf unseren Auftrag Nr. …
On unpacking the cases our Incoming Goods Control discovered that 15 items are missing.	Beim Auspacken der Kisten stellte unsere Warenannahme fest, dass 15 Positionen fehlen.
We are afraid that several units are - seriously damaged / defective. - broken / badly scratched / stained.	Leider sind mehrere Teile - schwer beschädigt / schadhaft. - zerbrochen / stark zerkratzt / verschmutzt.
We are sorry to point out that the repair work has been poorly executed.	Wir müssen leider darauf hinweisen, dass die Reparatur schlecht ausgeführt wurde.
We believe the damage may be due to rough handling in transit.	Wir glauben, dass der Schaden auf unsachgemäße Behandlung beim Transport zurückzuführen ist.
This is very inconvenient for us because …	Dies kommt uns sehr ungelegen, da …
We would ask you to - replace the faulty goods at your expense. - grant us a price reduction of 10%. - cut the price to €550.	Wir möchten Sie bitten, - die mangelhafte Ware auf Ihre Kosten zu ersetzen. - uns einen Preisnachlass von 10% zu gewähren. - den Preis auf €550 zu senken.
We will keep the damaged goods until we hear from you.	Wir werden die beschädigten Waren behalten, bis wir von Ihnen hören.
Responding to a complaint	**Auf eine Beschwerde / Reklamation antworten**
Thank you for your email drawing a serious problem to our attention.	Danke für Ihre E-Mail, mit der Sie uns auf ein ernstes Problem aufmerksam gemacht haben.
We wish to apologise for this mistake.	Wir bitten für diesen Fehler um Entschuldigung.
We are extremely sorry for the poor service you have received.	Es tut uns außerordentlich leid, dass Sie einen unzureichenden Service erhalten haben.
We will investigate the matter thoroughly and inform you of the steps taken.	Wir werden die Angelegenheit gründlich untersuchen und Sie über die Schritte informieren, die wir unternommen haben.
Please return the faulty items at our expense.	Bitte senden Sie die mangelhaften Artikel auf unsere Kosten zurück.
We are prepared to reduce the price by 10% if you decide to keep the goods.	Wir sind bereit, den Preis um 10% zu senken, wenn Sie sich entschließen, die Ware zu behalten.
We hope that this proposal will find your approval.	Wir hoffen, dieser Vorschlag findet Ihre Zustimmung.

Audioscripts

◎ A 4.1 Module 1 Introduction to vehicles, A 3, 4

Radio host: Welcome to today's radio programme. My guest today is Mr Gibbs. Mr Gibbs is a local car dealer. He is here to give us some information about the buying behaviour of people when looking for a new car. Mr Gibbs, would you please start by telling us something about the most popular type of car people buy nowadays?

Mr Gibbs: Of course. Well, most people, particularly young families, buy MPVs. Let me explain what is meant by an MPV: MPV is the abbreviation for 'multi-purpose vehicle'. Buyers prefer the roomy and versatile cabin; usually it offers seating for up to seven people. Typically, these cars have low running costs, are easy to drive and have a large boot. Some models offer plenty of space, others are a bit smaller, but they offer one of the best handling in their class.

Radio host: So what about four-wheel drives? These are still very popular, right? You certainly see a lot of them on the road.

Mr Gibbs: Sure, 4x4s are still quite popular. Interestingly though, most of the people who buy a 4x4 will never drive off-road. People usually buy them because they want a comfortable, spacious and safe vehicle. And that's the reason why people prefer crossovers. These cars are usually as spacious as a regular estate car but look like a 4x4. Some of them are equipped with four-wheel drive, and some have just front wheel drive. And guess what: most of these cars are also never used off-road. You sit high up in these vehicles and a lot of car owners like having a better overview of the roads – and they like the design.

Radio host: Thank you very much for explaining the most important features. But what about those people who are interested in smaller but more powerful cars?

Mr Gibbs: Well, these people favour a different type of automobile. Coupés are still very popular. They have just two doors, sometimes offer strong performance, and are bought chiefly because of their design. If you need more power, then you could go for a performance car, or a hot hatch. These cars offer very powerful engines, great performance, amazing grip on corners and a sporty design. The only problem is that these cars are generally much more expensive and not very family-friendly.

Radio host: Mr Gibbs, thank you very much for coming today and sharing your knowledge with us.

◎ A 1.1 Module A Meeting people, A 1, 2

Dialogue 1

Mr Neureuther: Neureuther, good morning.

Ms Hornbach: Good morning, Mr Neureuther.

Mr Neureuther: Good morning, Ms Hornbach.

Ms Hornbach: Mr Neureuther, the young man you were expecting from Denmark is here.

Mr Neureuther: Ah yes, please send him in. And Ms Hornbach, give Mr Dachs a call, will you please?

Ms Hornbach: I've already called him.

Christian: Hello, my name is Christian Eriksen. I come from Fanö in Denmark, and, uh, I'd like to work here as an intern for the next three months.

Mr Neureuther: Felix Neureuther. I'm the manager of the company.

Christian: Nice to meet you, Mr Neureuther.

Mr Neureuther: Nice to meet you too, Christian. I hope you had a pleasant trip.

Christian: Yes, it was fine, thank you. It's not far from Fanö to Kiel.

Mr Neureuther: Fine. Now, let me see, I guess you've handed all the necessary papers over to my secretary, Ms Hornbach?

Christian: Yes, I have.

Mr Neureuther: Oh good, Mr Dachs. I don't believe you've met Christian Eriksen. He'll be our new intern for the next three months. Christian, this is Mr Dachs, our personnel manager.

Mr Dachs: Nice to meet you Christian.

Christian: Nice to meet you too, Mr Dachs.

Mr Neureuther: Christian, Mr Dachs will take you to your team, so you can get to know them all. See you later, Christian.

Christian: See you later, Mr Neureuther.

Dialogue 2

Mr Dachs: Morning, everybody.

All of them: Good morning, Mr Dachs.

Mr Dachs: I'd like you to meet our new colleague. He'll be working with us for the next three months. He's in his second year of training. Christian, would you like to introduce yourself?

Christian: Hello, my name is Christian Eriksen. I come from Fanö, that's an island off the coast of Denmark. The company I work for is a bit smaller than yours; there's just the boss, a skilled worker and two trainees.

Silke: Hi Christian, nice to meet you. My name is Silke, and I'm also in my second year of training.

Christian: Nice to meet you too, Silke. I'm glad you speak English, because my German is not so fluent. But I'm doing my best to improve.

Silke: No problem, we'll manage.

Josef: Hello Christian, I'm Josef, but everybody calls me Joe. I'm your foreman. My English is not perfect, but Silke can always help. I've been here for quite a long time. Isn't that right, Mr Dachs?

Mr Dachs: Yes, Joe has been with us for over 20 years now. I think for the first few weeks you'll work with him. Well, let's get back to work now. If you have any questions, you can ask Joe, but you can always come to my office, too.

Christian: Thank you, Mr Dachs.

⊙ A 4.2 Module 2 Tools in a car workshop, C 2

Welcome to Milestone's Automotive Tools!
At Milestone's Automotive Tools we sell products in all major tool categories including automotive, construction, DIY, gardening, assembly and engineering. We have recently diversified into hobby, leisure and other non-tool lines.
Our company is renowned for the high quality of its products and has received awards from well-known consumer magazines. Milestone's Automotive Tools is a family-run company and has been supplying hand tools to the industry since 1923. We are located in Northumberland, England, and have been here for over 30 years.
Our catalogue is full of sensational new products. If you would like a copy, please register online and write us an email.

⊙ A 1.2 Module B Your company, A 4

Herr Kirchner: OK, so we've covered human resources, accounting, sales and marketing, and purchasing. Now we come to the more technical side of the company, and – of course – the area where you will be spending your time. OK, do you have any questions before we continue?

Sean: Yes, just a general question really. How old is the company?

Herr Kirchner: Older than you, Sean, that's for sure! Krone Elektronik was founded by Dr Florian Krone in 1954, and is still a family-owned company to this day.

Sean: And is this the headquarters here in Bochum?

Herr Kirchner: Yes, it is. Krone has nine manufacturing plants and sales offices across Europe, but this is the largest. In fact, we also have two in the USA – in Chicago and Baltimore.

Sean: Are there any plants in Great Britain?

Herr Kirchner: No, there aren't. But we are opening a new factory in South Korea next year.

Sean: And how many people work for Krone Elektronik?

Herr Kirchner: That's a good question, Sean. I don't know the exact number, but it is at least 10,000, if you include the USA. OK, let's go on with the tour, shall we? The building to your right is our research and development department. They are currently working on a new braking system for a major German car company.

Sean: And what goes on in the large building next to it?

Herr Kirchner: This is our main building, which has three functions. Firstly, it is where the production planners work – and this is where you will spend most of your time. Their job is to optimize the manufacturing process by keeping production times and costs down. Secondly, as you can see, here are also the production halls, where you will start your internship.

Sean: You mentioned three functions. So what else happens over there? It's a really big place.

Herr Kirchner: Yes, it's over 6,000 square metres. Those people you can see over there are responsible for quality control. They test all of the components and devices before they go to the customer.

Sean: How many components are made here in Bochum?

Herr Kirchner: Well, one of the main components we produce here are the cylinder heads, and we make thousands of them every year. And each and every one of them must be inspected by the guys in quality control.

Sean: I guess that's an important job. I'm really looking forward to getting to work.

Herr Kirchner: That's great. I will introduce you to the foreman this afternoon. OK, finally, behind this building we have the IT department. At the moment, they are very busy dealing with the problems caused by our new operating system.

Sean: Oh, that doesn't sound too good. What exactly is the problem?

Herr Kirchner: I'm not sure exactly, some kind of interface difficulties, I think. OK, let's go into the canteen and have some lunch, shall we? Then I can answer some more of your questions.

⦾ A 4.3 Module 3 The body shop, B 5, 6

Interviewer: The car body is put together in the body shop using different joining technologies. How do you join the parts? I mean, how many different joining methods are there?

Bernd: Well, there are several joining technologies. The most common technology used in the body shop is spot welding, also called resistance spot welding. That is when the two pieces of sheet metal are melted to join them together. An innovation in welding technology is when you use laser to weld the parts together. It's the same principle: you melt the parts and fuse them together, but it's much quicker – you need a lot less time – in fact it's three times as fast. And then of course, in some cases you have to join different materials like aluminium and steel together. These two materials cannot physically be welded together, so we have to use mechanical joining methods such as rivet welding or adhesive bonding.

Interviewer: What is the difference between rivet welding and adhesive bonding?

Bernd: Rivet welding is basically when a self-piercing rivet is stamped through the two plates that are to be joined and the rivets crimp to clamp the plates together. Adhesive bonding is basically when you just glue the two parts together. In those cases when we need to have a very stiff or rigid body, we often use a combination of both rivets and adhesive.

Interviewer: What are the most common materials used in passenger car production today?

Bernd: Our goal is to build a stiff and light body. To achieve this we use a combination of different materials. We use a variety of different steels like cold-formed regular steel and hot-formed high-strength steel. Aluminium is often used for the hang-on parts because it is lighter. However, it is more expensive and so a lot of cars are still built with steel hang-on parts. And then of course there are also materials we use like thermo plastics or sheet moulded compounds (SMC) for hang-on parts or fibre reinforced plastics which may increasingly be used for structural components in the frame in the future; for example the side member. It's really a bit like cherry-picking, using the most suitable material for each part.

Interviewer: What passive safety issues are considered when choosing materials?

Bernd: Well, in a normal saloon the B-pillar has to be very strong and rigid to protect the passengers if there is a side impact crash. So that is a typical place for hot-formed steel. In a convertible we also use hot-formed steel for the A-pillar to ensure roll-over protection. Aluminium is the lightest metal and it is often used for the hang-on parts.

Interviewer: How is quality assured in the body shop?

Bernd: To ensure high and stable quality we have an online monitoring system for all the joining technology processes. That means that every single welding spot is continuously checked. The second quality loop is when, at regular intervals, the whole body of a car is destroyed in order to check the strength of every single joint.

Interviewer: What are typical quality issues in the body shop?

Bernd: One of the most challenging tasks is to make sure the hang-on parts fit well. For example the gaps between the door and the frame must be the same. The door must not protrude anywhere – it has to be flush.

Interviewer: How do you think the body shop will change in the future?

Bernd: Today, most automated body shops are planned to produce one car family, so all types of one model. What we are working on now, is being able to build several body types in one and the same body shop – so a body shop that can adjust itself to a new type of body. Time to market will be more important in the future and we have to be able to adapt to smaller volume production. Those days are gone when we could build the same number of a particular model for eight years.

A 1.3 Module C Telephoning, B 5, 6

Alex: Fertighaus Henke, Alex Bauder.

Caller: Yes, hello. Do you speak English?

Alex: Er, yes, a little. How can I help you?

Caller: Well, my name is Matthew Reid and your company is currently building my house. I'm a little confused. I thought this was Mr Deschler's telephone number. He is the construction manager, isn't he?

Alex: Yes, he is, but unfortunately he is off sick today, so all of his calls are being diverted to me. Can I help you with something?

Caller: Well, I was calling to give Mr Deschler some changes to our house.

Alex: OK, and what was your name again, please?

Caller: Reid, Matthew Reid.

Alex: Could you possibly spell that please?

Caller: Certainly. It's R-E-I-D.

Alex: R-I-E-D.

Caller: No, R-E for England, I for India D.

Alex: Sorry, R-E-I-D. OK, Mr Reid, and what is the address of the house?

Caller: Well, so far it is only a foundation pit, but it is Ringstraße 26 in Lernsdorf.

Alex: Ringstraße 26, Lernsdorf. Ah, yes, I know the site. We are scheduled to begin on your cellar next week, I think. OK, if you give me the information, I will pass it on to Mr Deschler. Let me just find a piece of paper … OK, I'm ready.

Caller: First of all, in the kitchen we need to move the extractor fan twenty centimetres to the left, so that it is directly over the hob. And the hole needs to have a diameter of one hundred and eighty-four millimetres, and we also need …

Alex: Sorry Mr Reid, but I am afraid any changes to your kitchen need to be discussed with the kitchen fitters. Are there any changes to the plans for any other part of the house?

Caller: Er, just a moment. Let me look at my list … OK, we need to change the height of the toilets in the bathroom. At the moment they are planned to be forty-five centimetres and we would like them to be fifty. Plus, we need to add another two hundred and twenty volt socket to the right of the bathroom mirror for my electric razor.

Alex: OK, anything else?

Caller: Yes, my wife and I have decided that we will need a larger carport. I have spoken to a friend and he says we should make it eight metres by six metres by three metres. Can you find out how much extra this would cost, please?

Alex: Of course. Let me just confirm, eight metres long by six metres wide. And how high should it be?

Caller: Three metres high. Of course, this means that we will have a smaller garden, but I think a carport is very useful. You can also ask Mr Deschler how big our garden will be with the new carport. It is currently one hundred and thirty-five square metres.

Alex: OK, I will ask him to adjust the plan and get in touch with you as soon as he is back in the office.

Caller: That's great. Oh, just one more thing. My wife wants to put larger windows in the living room. At the moment, the windows in the plan are one hundred and forty-eight centimetres by one hundred and eleven. She would like them to be one hundred and eighty-six by two hundred and twenty-two.

Alex: One eight six by two two two. OK Mr Reid, I will pass this information on to Mr Deschler and ask him to give you a call. Does he have your telephone number?

Caller: Yes, I think so, but let me give you my work number, too. It is 0391 587660…. Thank you for your help Mr Bauder. Goodbye.

Alex: You're welcome Mr Reid. Goodbye.

A 4.4 Module 4 The paint shop, B 3, 4

Mr White: Welcome to the second part of the painting process. The area we are now about to enter has to be absolutely dust-free. Even the smallest impurity can ruin the paintwork.

Michael: How exactly can a small piece of dust ruin the paint?

Mr White: Well, you would be able to see the blemish on the shiny surface. If you ran your hand across the surface, you would also be able to feel it.

Michael: I see.

Mr White: Everyone who works in this area has to wear special protective clothing. Everything has to be covered so they wear hoods, face masks, gloves, boots and overalls. Before we enter this area, we have to do the same. So, would you please first put on these overalls and make sure your hair is covered? Then please also put these plastic caps over your shoes.

Anna: Yes, of course.

Michael: Wow, I feel like an astronaut.

Mr White: Now we all look like astronauts. To enter this area of the paint shop we have to go through this airlock chamber here – one at a time. When you enter, the door behind you will close. You will remain in the chamber for a minute. During that time, you'll get blown about a bit. An air shower will blow off any dust or dirt particles. Then the door on the other side will open for you to pass through. Who would like to go through first?

Michael: I'll go first.

Mr White: Good, Michael is through ... now it's your turn Anna.

Anna: Right – here we go.

Mr White: OK, we are all in the dust-free zone. This is where the body shell gets its colour. The very latest special clean room technology has been used to create this area. That means that before the body shell can enter, this area has to be cleaned with ostrich feathers. And if any particles of dust are still on the surface after that, they will be blown off as the car body passes through the air blower as it enters the painting area over there.

Anna: There aren't many people here. How are the car bodies painted?

Mr White: People used to paint the cars, but today nearly all the painting is done by robots. Robots have become so agile that they can access and paint every single area of the body and we have fully automated production lines in the paint shop. The people you see are specially trained workers. They check the paint process after every stage.

Michael: What kind of paint is used?

Mr White: Today mostly water-based paints are used for the primer and base coat. This is because they are more environmentally friendly than the solvent-based conventional paints used years ago. You may be surprised to see how little waste there is – that is because high voltage is used for this to achieve the highest possible degree of transfer efficiency. At the moment that is over 70%, which is a lot! This coating process is called electric static painting. After the car body has been given its colour coat, it passes through an oven in a kind of pre-drying phase. This phase only lasts five minutes, the temperature of the oven is 85°C and when it is finished the body has a dry, matt surface. It is now ready for the top coat of paint – the clear coat. The clear coat is a solvent-based coat which both protects and gives the paint its ultra glossy finish. It is applied in the same way as the base coat and dried for 20 minutes at 140°C. Then a surface inspection is carried out and if necessary the surface is polished.

Anna: Is that the end of the painting process?

Mr White: No, not quite. The very final stage here in the paint shop is when wax is injected into the cavities using very fine nozzles to give the body its final corrosion protection.

Anna: Where does the body go from here?

Mr White: The next stage of production is the assembly. However, before each and every car body leaves the paint shop, it is carefully checked by experts. That way only perfectly coated bodies leave the paint shop. So that brings us to the end of our little tour of the paint shop. I hope you have enjoyed it and found it useful. Now, let's go out and take off all this protective clothing.

⦾ A 4.5 Module 5 Chassis and drivetrain, A 3

Paul: Can you tell me what exactly the chassis of a car is? It says different things in different books and magazines.

Supervisor: Yes, that's because technology has moved on since the first chassis was developed. But let's start at the beginning. In the traditional sense the chassis is basically everything except the body.

Paul: And the body is almost everything you can see when you look at the car, isn't it?

Supervisor: Exactly. That means the chassis is everything underneath the body. It is the frame of the bottom of the car and has all the parts that enable the car to work attached to it.

Paul: You mean the axles and wheels?

Supervisor: Yes, both the front axle, rear axle and wheels form part of the chassis. Also the suspension system and the steering system. We call that the 'body-on-frame construction' and it is still used for trucks and SUVs today.

Paul: So the body-on-frame construction is still used for trucks and SUVs. Why not for passenger cars?

Supervisor: In passenger cars we now talk about 'unit-body construction'. We now have different materials and stronger alloys so the body of the car is much stronger. Remember, a great deal has been done to strengthen the body.

Paul: What is a suspension system?

Supervisor: Basically everything that helps you steer the car, even out the bumps in the road and make the ride smooth. The shock absorbers form a large part of the suspension system.

Paul: Is a unit-body car nomally stiffer than the body on a full-framed vehicle?

Supervisor: Yes, it is – and that is why we now include the body and hang-on parts when we talk about the chassis!

Paul: Right, I see.

Supervisor: And of course that also means it is lighter!

Paul: OK, and that means it uses less fuel, right?

Supervisor: Exactly.

A 1.4 Module E Applications, C 1

Andrea: Hi David, how are you doing?

David: Hi Andrea. I'm fine thanks. How are you?

Andrea: Yeah, I'm OK. Busy, but fine.

David: I heard that you were thinking about applying for a three-month internship in the Northampton branch. Is that true?

Andrea: Yes, that's right. I really want to spend some time in England. I guess it was the same with you and coming to Germany.

David: Exactly. I wanted to see how things worked over here. So, you want to spend some time with my old colleagues, eh? That's great.

Andrea: Well actually, I wanted to ask you for a bit of advice if that's OK with you?

David: Sure, what can I help you with?

Andrea: Well, I have to officially apply for the internship, which means a letter and a CV.

David: OK, would you like me to read your letter?

Andrea: Yes, that would be great. I am going to finish it this weekend, and then I will bring it to work on Monday. Thanks. But where I really need your advice is with the CV. I was wondering if you had any tips for me. I know there are some differences between German and English CVs.

David: Yes, there are, you're right. Hmm … tips … let me see. OK, the first thing to remember is that your CV should be no longer than two pages.

Andrea: Two pages, OK.

David: Also, make sure you use good-quality paper and never send your CV without a covering letter.

Andrea: OK …

David: And don't forget to highlight your strengths and achievements. And when you talk about your work experience, start with your most recent job.

Andrea: Do I need to include a picture?

David: In England, a CV does not have a picture, but you can have one if you like. What is important is that you include the name and contact details of two references. These could be a supervisor or maybe a teacher.

Andrea: OK, great. Anything else?

David: Yeah, make sure that everything you write in your CV is the truth. If you are asked to explain something during the interview and you can't, then it could be pretty embarrassing. Be confident and sell yourself, but don't lie. Er … once you have finished the CV, read through it and check the spelling and the grammar. The best thing to do is to give it to a friend. And, when you send it, keep a copy for yourself to read before the interview.

Andrea: Thanks a lot for the advice, David. I'm really grateful.

David: Don't mention it, Andrea. Let me know if there is anything else I can do.

Andrea: Well … you did say somebody should read over it and check everything – and who'd be better than a native speaker …?

A 4.6 Module 6 Assembly and logistics, A 2

Trainee: What exactly happens in the assembly shop?

Mr Beck: The painted car bodies enter the assembly shop on a conveyor system and have all the selected parts added before they can finally be driven out as our perfect, high quality vehicles.

Trainee: Does that mean all the parts are added to the vehicle while it's moving along the assembly line?

Mr Beck: Not in small series production. But yes, definitely with mass production vehicles where time and costs are extremely important. The car moves from work station to work station along an overhead conveyor system at a pre-defined speed. Of course not all car manufacturers have exactly the same process, but on the whole they are quite similar. Would you like me to go through a typical process with you?

Trainee: Yes, please.

Mr Beck: Well, first we give the car its identity. We stamp a unique number onto the body of the car. This is called the chassis number; it remains with the car for its complete life-cycle. It has to be written on all the vehicle documents. Once that has been done the doors are removed.

Trainee: Yes, why do you remove the doors? Is it because the mechanics then have more room to install all the parts?

Mr Beck: Yes, that is one answer. And remember, a door is very complex today – it has electrically powered windows and airbags, etc. So the doors are taken to another line, the door trim line. This is where all the electrical parts of the door are added, the cables are laid, the door handles and arm rests attached, and the airbags installed and so on.

Trainee: So there are several lines, not only one?

Mr Beck: Yes, that's right. We have three main assembly or production lines: one for the interior of the car, one for the chassis and then one for the final assembly. Apart from the main lines, there are several smaller sub-assembly lines. For example the cockpit is pre-assembled on a separate sub-assembly line. So, all the parts of the cockpit are pre-assembled there, and then later the assembled cockpit is brought to the car interior assembly line and installed in the car.

ⓞ A 4.7 Module 6 Assembly and logistics, A 3

Trainee: So, what happens after the car receives the chassis number and the doors have been removed?

Mr Beck: Oh, yes. Well, once the doors have been removed, the sunroof is installed in the roof of the car. After that the wiring harness, that is all the cables are laid. When that's finished, all the electrical parts are attached.

Trainee: I see.

Mr Beck: Then, once the battery has been put in, the cockpit arrives from pre-assembly and is installed. The next step is to carefully fit the headliner.

Trainee: Are all the jobs performed manually?

Mr Beck: Most of these jobs are done manually. But the next task is to put in and seal the windshield and rear window. High-precision robots do that task.

Trainee: That must be amazing to see.

Mr Beck: Yes, it is quite impressive. I'll show you later. The windshield is heavy and has to be positioned very exactly. That used to be a difficult job for mechanics. So ... then after the robots have finished, it's time for the carpets to be laid. Then the side panels and edge protection are put in and the car moves on to the second line, the chassis assembly line, and the mechanics install all the piping to the underbody.

Trainee: Is the piping fitted to the bottom of the underbody?

Mr Beck: Yes it is, but remember the car is still hoisted at this stage so that the mechanics can stand under the vehicle. As you can imagine, that's a very tiring job. Then the petrol tank is assembled together with the fuel gauge unit and petrol pump. Once that has been completed, the most exciting part of assembly takes place – the marriage. The marriage is when the pre-assembled engine, the heart of the car, is mounted and the whole powertrain is raised and the upper body lowered. After the marriage the car is transported on a different conveyor system. This one is lower and the vehicle is almost at the same height as it is after the wheels have been attached. Then the pre-assembled front module, that is the bumper, cooling system, radiator grille and lights, is attached. After that the seats are installed and adjusted. When the seats have been installed, the wheels are attached to the axles and the car can stand freely – for the first time on its own.

Trainee: What about the doors? They are still missing, aren't they?

Mr Beck: Yes, you' re right. Fitting the doors is the next step and the fully assembled doors, complete with rubber sealant, are already waiting to be returned to the car they originally came from.

Trainee: Is it really necessary to return the doors to the car they originally came from?

Mr Beck: Yes, that is extremely important. Not only because of the colour and size. They must be returned to the original vehicle as only this will ensure the best fit quality. The next step is to pour in the liquids – so the water, oil, coolant and petrol, etc. – are poured into the radiator, the brake, power steering and the windscreen wiper tank. And then the final door fit is checked.

Trainee: Is that the end of the chassis line?

Mr Beck: Yes, it is.

Trainee: And what is there left to assemble? I mean what happens on the final assembly line?

Mr Beck: The final assembly line is really a kind of inspection line. So, first the lights are adjusted, the wheels are aligned and then the rolls test is carried out. This is when a computer tests the electronic functions in driving situations that are computer simulated. After that the engine and painted surfaces are checked and finally a rain check is carried out. That is when the car is put under a shower for quite a long time to make sure that no water gets into the vehicle and it is perfectly sealed. Once everything has been checked, the finished vehicle is ready to be shipped to the customer.

◎ A 4.8 Module 6 Assembly and logistics, C 1, 2

Visitor: Good morning Ms Thomas. Thank you for showing me around the assembly line – you certainly seem to have a very efficient logistics system in place.

Ms Thomas: Yes, we're very proud of what we have achieved. We used to have problems with forklift drivers. They complained they had too much work. There were often delays at the line and we had a few accidents.

Visitor: Really, what did you do? Did you reduce the amount of work or did you employ more people?

Ms Thomas: Neither. We realised we had an uneven workflow – so there were times when people really had too much to do and there were also times when they just stood around waiting.

Visitor: What were the causes?

Ms Thomas: The biggest problem was at the beginning – when the trailers arrived and were unloaded onto the docks. At certain times of the day, first thing in the morning for example, so many trucks arrived at the same time that we had traffic jams in the plant and everyone was busy trying to get the trailers unloaded. A few hours later, there were no trucks and the forklift drivers had nothing to do. So that was the first thing we changed, we now have trucks arriving continuously and as a result we have a more even workload at the docks and the forklift drivers are not rushed at any time.

Visitor: Sounds so simple! What else did you do?

Ms Thomas: Well, then we noticed we were storing the same parts in several different places. That meant increased handling. When we needed the parts on the line, we sometimes had to look in as many as eight different places. So we reduced the number of parts we have in stock.

Visitor: So you could put the same parts in one place?

Ms Thomas: Basically, yes. But we still didn't have enough space for incoming goods. We had too much stock because the inventory policies were not being followed correctly. To solve that problem we reviewed the inventory system and installed minimum and maximum signs to control the inventory levels. That way we could easily see when we had too many or too few parts.

Visitor: Did that also help stop the delays you mentioned?

Ms Thomas: No, not totally, anyway. The delays happened because the aisles were too narrow and the forklifts couldn't pass each other. We had traffic jams if you like! So we widened the aisles and that did the trick.

Visitor: Did that also reduce the number of accidents you had?

Ms Thomas: A bit, yes. But the thing that really reduced the number of accidents was the fact that we were able to reduce congestion by reorganising the loading routes – we introduced a kind of one-way system. Now the forklifts have set routes at set times and, as you saw, no congestion.

◎ A 1.5 Module F Socialising, A 1

Fabian: Hi, I'm Fabian.

Jan: Hello Fabian, my name's Jan. I'm with the group from Warsaw, Poland.

Fabian: Welcome to our company. Is this your first visit to Germany?

Jan: Actually, I've been here before. But my German isn't very good.

Fabian: Don't worry, we can talk in English. I should really practice more anyway. How was your trip? Was it a long journey?

Jan: No, not really. We took the plane from Warsaw to Stuttgart. Then someone from your company met us at the airport.

Fabian: Jan, would you like something to drink: coffee, juice, lemonade?

Jan: Sure, a lemonade would be good.

Fabian: OK, let's go to our canteen and sit down for a while. Here we go; I hope it's cold enough.

Jan: Thanks. So, I guess that you are also a trainee. What year are you in?

Fabian: I'm in my third year. What about you?

Jan: I've just started. How many trainees are there in your company?

Fabian: Er ... about ten, I think.

Jan: That's quite a lot. There are only three trainees in our company, all in metal engineering. I think your company is much bigger than ours.

Fabian: Yeah, we even have our own football team. We don't play in a league; it's just for fun.

Jan: That sounds great! You should visit us in Poland one day. Maybe we could arrange a match.

Fabian: That's a good idea. Are you a football fan?

Jan: Yeah. I don't play myself, but I like to watch.

Fabian: Which team do you like?

Jan: Manchester United.

Fabian: Really? They're my favourite team, too! Well, apart from VFB Stuttgart, of course.

Jan: Do you live in Stuttgart?

Fabian: No, I live in a small town nearby. I have to take the bus every morning. But when the weather's good, I can take my motorbike.

Jan: Cool! Unfortunately I don't have a motorbike. I get a lift from a colleague who works in the same department as me.

Fabian: What do you do in your free time?

Jan: I like going to clubs, and sometimes I go fishing with my friends. What about you? Playing football, I suppose.

Fabian: Yeah, football and riding my motorbike.

A 4.9 Module 7 Engine basics, A 3

Pat: So, all combustion engines have pistons in them.

Tom: Yes, Pat, that's right. However, all engines are different and you'll see they are not always positioned in the same way.

Pat: What do you mean?

Tom: Well, for example in some engines the cylinders and pistons are vertical and all arranged in one straight line. We call that inline. On the other hand in some engines they are positioned differently. Sometimes they are flat or what we call 'horizontally opposed'. That means they are horizontal, and there are two rows opposite each other. This arrangement is also known as boxer.

Pat: OK, so there are boxer engines – where the cylinders with the pistons are flat and in two rows – or otherwise they are in one row and are vertical.

Tom: Exactly. But we also have the V engines. The V8 engine has eight cylinders with pistons and they are arranged in two lines of four. The two lines are at an angle, a bit like a big letter V.

Pat: Oh right. So that means a V6 engine has six cylinders, with six pistons in two lines of three.

Tom: Yes, you've got it.

A 4.10 Module 7 Engine basics, B 1

Engineer: Good morning, welcome to the motor show. Is there anything I can do for you?

Visitor: Good morning. Yes, actually there is. I want to buy a new car and I'm not sure what to buy. Could you please tell me a bit about the difference between diesel and petrol engines? With the price of petrol going up all the time, what should people buy today? Are diesel engines still more fuel-efficient than petrol engines?

Engineer: Well, petrol engines have become much more efficient over the past years, especially since most car manufacturers have introduced electronic engines and fuel management systems. But you still get 30% less mileage out of a petrol engine than you do from a diesel engine of the same size.

Visitor: So what is the difference between a diesel and a petrol engine?

Engineer: Well, first of all diesel is thicker, it has more oil in it and it is not refined. That is one reason it is cheaper. Then of course there is the technology. To put it simply, the biggest difference between diesel and petrol engines is the way the explosions happen. In a petrol engine, the mixture of fuel and air is compressed in the combustion chamber and then ignited by sparks from a spark plug. In a diesel engine, on the other hand, the air is compressed first and then the fuel is injected. The pressure in the combustion chamber of the diesel engine can be twice as high as in the petrol engine and the air reaches very high temperatures. It gets so hot that when the fuel is injected into the combustion chamber, it self-ignites.

Visitor: So, basically the main difference is that petrol engines use sparks to ignite the fuel-air mixture in the cylinder, while a diesel engine compresses the air until it gets so hot that it can ignite the diesel as soon as it enters the combustion chamber.

Engineer: Exactly. Some diesel engines have glow plugs to help heat the combustion chamber a bit before you start the car. If you don't do that, the air might not get hot enough to ignite the diesel.

Visitor: Oh, yes. I remember my grandfather always used to press the glow plug before he started the car.

Engineer: Yeah – all diesel cars used to have glow plugs, but now only older ones or smaller ones, less highly developed engines have them. Larger engines have electronic sensors that measure the temperature and if it is cold, the fuel is injected a little later and the air in the cylinder is compressed a bit more, which of course means it gets hotter and that helps start the engine.

Visitor: Can you use diesel in a petrol car?

Engineer: No you can't. As I said, diesel is thicker and contains more oil, so it doesn't evaporate very well. The spark plug wouldn't be able to ignite it. The car wouldn't start and you would have to drain the entire diesel out before starting the car again.

Visitor: So I really must make sure I don't use the wrong fuel! I'll remember that next time I am at a petrol station! So which car is more expensive to buy?

Engineer: Well, the diesel engine is more expensive to produce, because it has to be much stronger to be able to withstand much higher pressure. So diesel-powered cars are more expensive to buy. But, if you are thinking of buying a new car, also think of your driving habits. If you drive a lot, then it is worth driving a diesel because you'll get back the extra cost with what you later save on fuel.

Visitor: Wow, now I understand a lot more. Thanks very much for all the information, it's been a great help.

Engineer: It was a pleasure. Here's my business card. If you have any more questions, give me a call.

Visitor: Thank you very much. I might do that. Goodbye.

⊙ A 1.6 Module G Presentations, C 1

Good morning ladies and gentlemen. Today I would like to present our new range of wooden furniture, Simply Wood. The range, which is shown here on the first slide, includes chairs, a large table, a lounger, and benches. I would like to start by talking about the materials used in the furniture. All of the Simply Wood range is made from single softwood timber and has been hand-crafted by our highly-skilled carpenters and joiners. If you look at this picture, you can see that the benches and the seats are designed with a low back, deep seats and long arms to offer a great deal of comfort and convenience. Thanks to a special varnish, the furniture is 100% weatherproof and will not suffer from discolouring. As you can see from this final slide, the furniture is also available in white, to give it a classic, timeless look. And do not worry, we have used a durable, white polyurethane coating that is suitable for every environment. On your handout you will see more details about each piece, including measurements. Thank you for your attention this morning. Are there any questions? Yes …

⊙ A 4.11 Module 8 Alternative engines, B 2

Bob: Hi, Frank. I'm going on a course next week to learn about the fuel cell car. I'm totally unprepared – I don't know a thing. Do you know anything about how the fuel cell car operates?

Frank: Don't worry. You'll find out about it there. A lot of people think it's the technology of the future. Basically the fuel cell is a type of electric car as it has an electric motor to propel the wheels.

Bob: So what is the difference between the two electric cars, then?

Frank: Well, as you know, the electric car gets its power from a set of batteries and when the batteries are empty, they have to be recharged. The fuel cell car also has a battery pack but it generates its own electricity.

Bob: How does it do that?

Frank: To put it very simply, the fuel cell car has a tank or rather several tanks, a bit like a petrol tank. But it doesn't store petrol, it stores hydrogen in the form of compressed gas.

Bob: Ah, so that's why it's also called the hydrogen car.

Frank: Exactly. It's the chemical reaction of hydrogen together with oxygen from the air that generates an electric current.

Bob: What about the lithium-ion batteries? Is that where the electricity from the fuel cell is stored?

Frank: Yes, it is. But the battery also stores the energy it gets from regenerative braking and supports during acceleration. You'll learn about that in your course.

Bob: Sounds pretty complicated to me! Are there any emmissions?

Frank: That is the unbelievable thing. Some people say the air is cleaner after you have driven the car than it was before because the only emission you have, is water.

Bob: Wow – and what happens when you run out of hydrogen?

Frank: That's exactly the same as when you run out of petrol. You drive to the filling station and refill the tank in a couple of minutes. It's as simple as that. You'll find out a lot more on your course. Have fun and let me know how it goes.

Bob: Thanks, sure. I'll send you an email.

A 4.12 Module 8 Alternative engines, C 3

Interviewer: Paul, how far have you got to finding the best new technology, and what do drivers today expect?

Paul Winter: Today's drivers want a vehicle which they can drive for at least 300 miles, that is nearly 500 kilometres, before they have to stop to refuel. They also want to be able to refuel the car quickly and last but not least they want to be able to drive relatively fast – or at least be able to keep up with other cars on the road.

Interviewer: But isn't that exactly what we already have with the car that has a traditional internal combustion engine?

Paul Winter: Exactly. You are absolutely right. Those are the biggest advantages of the petrol-driven car. The petrol car does all that and is also cheaper than all the other technologies on the market today. However, despite large efforts to reduce fuel consumption, it still burns too much fuel and so the emissions are too high. Those are the two biggest factors that speak against the petrol car – we all know that one day we will run out of petrol and air pollution has to be reduced to save our planet.

Interviewer: So what kind of cars do you think we will be driving in the future?

Paul Winter: Well, as you know, we are working on three major technologies: the all electric car that runs on batteries, the fuell cell and the hybrid. These all have their advantages but also their disadvantages.

Interviewer: So what exactly are the pros and cons of each kind of technology?

Paul Winter: Let's take the electric car. The biggest thing that speaks for that technology is the fact that it has absolutely no emissions. It is also very quiet, which is an advantage. However, it is a disadvantage at the same time. Just think of the pedestrians – they are used to hearing a car approach; if a car approaches silently, they may not notice it and have an accident. Then of course the biggest disadvantage is that it takes a long time to recharge the batteries. Even using a speed charger takes around two hours and there aren't very many cities that have speed chargers at all. And if they do, there aren't nearly enough of them. On top of that, the batteries are heavy and big, so they take up a large part of the car. Another disadvantage is that they can't be driven very far – the maximum distance is around 100 miles or 160 kilometres.

Interviewer: Does the battery last forever?

Paul Winter: Good point – I'd forgotten to mention that. A battery doesn't last forever. After a time it needs to be replaced and that is very expensive. Batteries also contain harmful substances and are not easy to dispose of. So when you throw them away, they are also not very good for the environment.

Interviewer: There seem to be a lot of disadvantages of electric vehicles. What about hybrids?

Paul Winter: The advantages of the hybrids are that fuel consumption is lower because they have very small engines to back up the electric engine. And although they also have batteries, in some hybrids the batteries don't need recharging.

Interviewer: And what about the cons?

Paul Winter: The biggest problem with the hybrid is weight. Remember: it has two engines, a generator and a few batteries – not quite as many as the electric car but it all adds up.

Interviewer: What about the costs?

Paul Winter: That is also a disadvantage. Hybrids are still quite a bit more expensive than petrol cars, but when you consider that they use a lot less fuel, then if you drive a lot, the savings that you have over the years on fuel costs may give you most of that money back.

Interviewer: So that leaves us with the fuel cell and isn't that even more expensive?

Paul Winter: Definitely. The fuel cell car is by far the most expensive. But it is a very exciting new technology. In years to come we may be lighting whole streets of houses with fuel cells. Fuel cells use hydrogen and air to generate their own electricity. Hydrogen is everywhere and is renewable. You have no emissions and no recharging of batteries.

Interviewer: That means there are a lot of advantages – so is the only disadvantage that it is so expensive?

Paul Winter: That's the main disadvantage – it really is just too expensive at the moment. Other disadvantages are that people are worried about driving a car with a tank of hydrogen. Hydrogen is highly flammable. But on the other hand so is petrol and we don't seem to be too worried about that. Of course we don't have enough hydrogen refuelling stations at the moment either. So that is another disadvantage today.

Interviewer: So does that mean the petrol-driven car will be discontinued?

Paul Winter: No, not in the immediate future anyway. We will continue to research and develop all of the technologies including the combustion engine. We have managed to find ways of reducing fuel consumption in the internal combustion engine dramatically over the last few years and will continue to do so.
So I think the petrol engine will continue to be the most popular car on the roads for quite some time to come.

A 1.7 Module H Dealing with customers, A 4

Answering machine: Guten Tag. Sie sind mit Stefan Bauder verbunden. Leider bin ich im Moment nicht erreichbar. Bitte hinterlassen Sie eine Nachricht nach dem Ton. Danke!

Brian: Hello Mr Bauder, this is Brian Jenkins calling from New Wave Marketing in Hamburg. I am calling about the new IT network that your technicians set up last week. Unfortunately, we have been experiencing some problems.
Since your technicians were here, some employees have had difficulty connecting to the Internet, and others have no access to our network at all and cannot even log on to their computers! However, the worst thing is that some important files and documents seem to have been lost as a result of the migration to the new software. One of these files was a presentation for a very important client. This is a particular problem for us because we have a meeting with this client on Monday, and we need the presentation and do not have enough time to redo it. Could you please look into this problem for us urgently as time is running out?
If we do not hear from you by the end of today, we will have to look elsewhere. I will be in meetings this afternoon, but you can contact me per email under B dot Jenkins at NWM...de. Thank you.

A 4.13 Module 9 Safety, A 6, 7

Interviewer: Teresa, you say a lot has been done to prevent SUVs and crossovers from rolling over. Does that mean they cannot roll over?

Teresa: No, that is definitely not what it means. Even with the best technology and a lot of testing we aren't always able to prevent a tall car from rolling over if the worst case sets in. That is, for example, when a car hits a curb or skids on ice. There is almost nothing we can do in those cases.

Interviewer: What happens then?

Teresa: That is when the in-cabin safety systems are triggered.

Interviewer: And what exactly does that imply?

Teresa: Well, all the seats have seat belt tensioners which automatically tighten and help keep the occupants in their seats. Then the airbags are deployed, that is the inflatable side curtains deploy to protect the heads of all the two or three rows of passengers. The curtains remain inflated for a few seconds to help prevent any passengers who aren't wearing their seat belts from being ejected through a broken window.

Interviewer: What is the difference between a normal airbag and the side curtain or window bag?

Teresa: The side curtain prevents the head from directly hitting the window or the B-pillar or C-pillar. It also protects the occupant from any objects that pierce the passenger cell, such as for example the branch of a tree or parts of another car. It can also protect passengers from additional risk caused by bits of glass or other flying objects.

Interviewer: What happens after the accident? Can the passengers get out of the car and is anyone notified?

Teresa: Quick rescue of injured passengers is absolutely vital. That is also part of the integrated safety concept. For example the doors are automatically unlocked and a message that gives the exact location of the accident can be send to an emergency call centre. And to make sure any remaining occupants are safe the ICRS – the intelligent crash response system – sets off all the hazard warning lights to warn any other vehicles.

A 4.14 Module 10 Wheels and brakes, A 4

Speaker 1: Why do I need snow tyres in winter? Do I really need snow tyres if I have a new set of all-season tyres?

Speaker 2: If you live in a country where the temperatures fall below 7° C, the answer is yes because that is when your all-season tyres start to lose their grip. If you live in an area where the temperatures drop to -5°C, then snow tyres are a must because your all-season tyres will be as good as useless.

Speaker 1: Why are snow tyres better in freezing conditions?

Speaker 2: The rubber compound used to make snow tyres is much softer and so the tread stays softer even at low temperatures. This helps maintain grip which affects the braking distance. Rubber on all-season tyres becomes very hard when the temperatures drop and this makes braking almost impossible.

Speaker 1: Is it really only the rubber compound which makes the difference?

Speaker 2: No, it isn't just the softer rubber which makes driving at low temperatures safer. It's also the deeper tread that the snow tyres have. They help you get through the snow.

Speaker 1: Well if snow tyres are so much better, can I just have snow tyres and drive them all year?

Speaker 2: Unfortunately not. The soft rubber will wear down quickly if you drive winter tyres in summer, so they won't last very long and again, braking will be more difficult.

Speaker 1: So, I suppose I will need two sets of tyres.

Speaker 2: Yes, and if you have them on two sets of rims, you'll see they are easy and quick to change.

A 4.15 Module 11 Car systems, A 3

Interviewer: Could you tell us what the cooling system of a car looks like and how it works?

Engineer: Sure. The cooling system is very important because if the engine overheats, it could cause irreparable damage. Let's look at the picture. You see this big flat tank here? Well that's the radiator there, behind the grille, and that is the most important part of the cooling system. The radiator is filled with a liquid which we call coolant. It's normally a mixture of antifreeze and distilled water.

Interviewer: Is that why coolant is often just called antifreeze?

Engineer: Yeah, exactly.

Interviewer: I see.

Engineer: So, the coolant flows through small channels inside the radiator and is cooled by the air that comes in through the grille while the car is driving. Right behind the radiator you see a kind of propellor – that is the fan. There are often two of those. They rotate when the engine is running and also help reduce the heat of the coolant.

Interviewer: Where is the engine?

Engineer: You see that big block? Well, that is the engine and between the engine and the fan, you can see the water pump. The water pump is attached to the radiator via the lower radiator hose. It draws cold coolant from the radiator through the hose and pumps it into the engine. The coolant travels through the engine and cools it. When the engine reaches its optimal temperature, the valve of the thermostat, that is this part here at the top of the engine, opens so that the hot coolant can flow into the upper radiator hose back into the radiator and be cooled again. That's basically it.

Interviewer: But there are still some parts you haven't mentioned.

Engineer: Yes, you're right. On top of the radiator here you can also see the pressure cap. The pressure cap keeps the pressure inside the radiator at a certain level to stop the coolant getting too hot. If the pressure gets too high, a valve in the cap opens, this releases the pressure and a small amount of coolant comes out. This small amount goes into the hose here and travels to the reserve tank. Can you see that? It's this small, clear plastic tank.

Interviewer: Yes. How long does the coolant stay there?

Engineer: Only until the engine has cooled down. As soon as the engine is cool, another valve draws the coolant out of the reserve tank again and sends it back into the radiator.

Interviewer: That is clever! What about this part at the top of the engine here with the two hoses coming out? What is that?

Engineer: Ah right. That is the heater core and those are the heater hoses. You need those when it's cold outside and you want to heat the interior of the car.

A 4.16 Module 11 Car systems, C 3

Let me tell you about some of the most common problems in the cooling system and what actions you can take to find out the cause. You may see some coolant on the ground underneath the engine of the car. Then the first thing you should do is check the hoses. See if you can find a small hole that coolant could be leaking from. If the hose is fine, then check the radiator itself. See if it has a crack. Another problem that you could have is that your engine may overheat. That is a serious problem and can have different causes: To begin with, check the level of the coolant in the radiator. If that's OK, check the water pump to make sure that it is not leaking in any way. If there are no problems there, the next step would be to check the thermostat. The engine should be warm and all you have to do is touch the upper radiator hose and make sure it's hot. If it is cold, then the thermostat is probably stuck in the closed position and no coolant can circulate. If no hot air at all comes in through the vents, check the heater core to see if it is blocked or even broken.

A 4.17 Module 12 Maintenance and repairs, C 1

John: Good morning, Happy Car Repairs, John speaking. How can I help you?

Ms Hall: Good morning, John. This is Mary Hall speaking. I'm calling to make an appointment to have my car serviced.

John: Is that just for an oil service or a full inspection service?

Ms Hall: I'm not sure, the display just said I need an oil service.

John: I see, what's the car registration number and I'll just check for you.

Ms Hall: Thank you. It's JY- KLM 605.

John: Right, one moment ... ah yes, here we are. Ms Hall I see your car is nearly three years old. That means it will be coming up for its first MOT test.

Ms Hall: Oh, where do I have to have that done?

John: That has to be done at an authorised test centre. But if you like, we can arrange to have it done for you. We'll take it there after we have done a full inspection service to make sure we check everything and do all the necessary repairs before the test. Then we'll know it will pass.

Ms Hall: Oh, that's wonderful. When can I bring it in?

John: How about next Thursday morning at 8:00?

Ms Hall: Next Thursday at 8:00 ... yes, that'll be fine.

John: Would you like me to book you a courtesy car? That way it won't cost you anything and you can drive to the office while your car is here.

Ms Hall: Yes, please John, that would be very kind of you.

John: Right, that's booked. So I'll see you next Thursday, then.

Ms Hall: Yes. Goodbye John.

John: Goodbye Ms Hall and thank you for calling.

A 4.18 Module 13 Trucks and lorries, B 5, 6

HLD: Good morning, Happy Lorry Driving, how can I help you?

John: Good morning, this is John Farlow speaking. I'm ringing because I'd like to know if I can get a driving licence for LGVs and I thought you might be able to help me.

HLD: Yes, certainly Mr Farlow. Do you already have a full passenger car driving licence?

John: Yes, I do.

HLD: Good. How long have you had it for?

John: I've had it for four years and I have never had an accident or anything.

HLD: That is excellent and means you are already allowed to operate any vehicle up to 3.5 tonnes.

John: Oh, I never realised that. What do I have to do to be able to drive heavier vehicles and combination vehicles? I mean articulated lorries with more than one trailer.

HLD: You need an extra licence for anything over 3.5 tonnes. That means you have to do a theory and hazard perception test first.

John: What happens in the hazard perception test?

HLD: In the hazard perception test you watch video scenes and have to click on the mouse button when you would react. So for example, you are driving along a straight road and see an obstacle – so you have to click on the mouse button when you would slow down. Things like that.

John: I see. That must be quite fun.

HLD: Yes it is. Anyway, as I was saying when you have passed the theory and hazard perception test, you can take the practical driving test. And finally you now also have to pass the Driver Certificate of Professional Competence – what we call the Driver CPC.

John: Do all drivers have to pass the Driver CPC?

HLD: Yes, that is, all drivers who drive a vehicle over 3.5 tonnes. In fact, all professional European LGV drivers then have to do a 35-hour CPC training every five years. It's a legal requirement.

John: Is there anything else I have to do?

HLD: Yes, you will also have to pass a health check.

John: Wow, that is quite strict.

HLD: Yes, I suppose it is, but driving a big truck like that is not as easy as you might think and you carry a big responsibility. It would be extremely serious if you had an accident with a passenger car because you have a heart problem for example.

John: So are there any extra requirements if I want to drive an articulated lorry?

HLD: If you wish to drive an articulated lorry, you have to have had a licence to drive a medium rigid or heavy rigid lorry for over 12 months.

John: Oh, so I can't get that licence immediately?

HLD: No, I'm afraid you can't. You need to have experience in driving a large rigid truck first.

John: That's a pity. What about driving a multi-combination lorry? I mean an articulated lorry with more than one trailer?

HLD: That is, of course the most difficult lorrry to drive. You will be able to take the test for a Multi Combination licence after you have had a Heavy Combination or a Heavy Rigid vehicle licence for more than one year. Then you have to go on a special training course for multi-combination and pass that.

John: Right. Well, thank you very much for all that information.

HLD: It was a pleasure talking to you. Would you like me to send you an email with the most important facts?

John: Yes, please. That would be great. I've forgotten most of what you told me already.

HLD: Yes, I can imagine. I'll get the email off to you this afternoon. If you have any more questions, just give me a call. Goodbye.

John: Thanks. Goodbye.

Grammar overview

Present Tense (Gegenwart)

Bildung: Present Tense (simple)		
Merke: Bei der 3. Person Einzahl *(he / she / it)* wird ein *s* an die Endung des Verbs angehängt.	I like / you like / he (she / it) likes we like / you like / they like	computer games.

Verwendung: Present Tense (simple)	
– bei allgemein bekannten Tatsachen und Aussagen	– Most cars run on petrol.
– bei regelmäßigen Handlungen: Schlüsselwörter: *usually, normally, always, never, sometimes, often …*	– I usually start work at 7 o'clock.
– bei Gewohnheiten	– He doesn't smoke.
– bei Fahrplänen, Öffnungszeiten etc. für die Zukunft	– My train leaves at 10 o'clock tomorrow morning.

Bildung: Present Tense (progressive)		
am / are / is + Infinitiv mit Endung *-ing*	I **am** / you **are** / he / she / it / **is** we **are** / you **are** / they **are**	working on a new project.

Verwendung: Present Tense (progressive)	
– bei Handlungen im Augenblick des Sprechens	– I am reading a manual at the moment.
– bei vorübergehenden Handlungen	– I'm staying at the Royal Excelsior Hotel.
– bei (fest) geplanten Handlungen in der Zukunft	– I'm meeting Ms Winter next week.
– bei gefühlsbetonten Aussagen mit *always* (negativ besetzt)	– He is always complaining about his boss.

Past Tense (Vergangenheit)

Bildung: Past Simple	
ed-Endung bei regelmäßigen Verben	I **worked** in Munich last year.
2. Form bei unregelmäßigen Verben *(1. meet / **2. met** / 3. met)*	He **met** Mr Miller yesterday.

Verwendung: Past Simple		
bei Vorgängen, die in der Vergangenheit abgeschlossen wurden; oft mit Zeitangabe. Signalwörter u. a.: *yesterday, in (1990), two weeks ago, last year, last month, last week.*	I finished my studies	in 2008. 3 years ago. last week.

Present Perfect (Vollendete Gegenwart)

Bildung: Present Perfect (simple)		
have oder *has* + *ed*-Endung (bei regelmäßigen Verben)	I / you / we / they **have** worked He / she **has** worked	as a trainee.
	It **has** worked well for us.	
3. Form bei unregelmäßigen Verben *(1. see / 2. saw / **3. seen**)*	I have **seen** the new car.	

Bildung: Present Perfect (progressive)			
have / has + *been* + Infinitiv mit *-ing*	I / you / we / they He / she	**have** **has**	**been living** in Munich for 5 years.

Verwendung: Present Perfect (simple and progressive)	
– bei Handlungen in der Vergangenheit **ohne Zeitangabe**	– I have written three letters.
– bei noch andauernden Zeitbestimmungen	– We haven't had any complaints so far / today …
– bei Handlungen, die in der Vergangenheit begonnen haben, aber noch nicht abgeschlossen sind, wobei ***for*** bei Zeiträumen *(two hours / three weeks)* und ***since*** bei Zeitpunkten *(2001 / January)* benutzt wird	– I have been living in Stuttgart **for** 20 years. – I have been living in Stuttgart **since** 1982.
– Einige Verben werden in der Regel nicht in der Verlaufsform verwendet.	– I have **had** the car for five years. – I have **known** the company since 2010.

Future (Zukunft)

Bildung und Verwendung	
Shall / will drückt äußere Umstände aus.	Most of this year's teleworking jobs **will** be offered in autumn.
Will drückt die Bereitschaft oder Entschlossenheit aus, etwas zu tun, sowie Versprechungen. ***Will*** wird auch benutzt für eine spontane Entscheidung.	I'**ll** do my best to get the job. Wait a minute, I'**ll** help you!
Going to wird verwendet, um persönliche Planungen und Vorhaben auszudrücken. ***Going to*** wird auch verwendet, wenn bereits Anzeichen dafür da sind, dass etwas eintreten wird.	I **am going to** apply for a job in teleworking. *(Ich habe vor…)* Look at those clouds! It'**s going to** rain.
Present continuous wird verwendet um eine feste Vereinbarung (gewöhnlich mit Zeitangabe) mitzuteilen.	Jennifer is **meeting** her colleague this evening.
Present simple wird für die Zukuft verwendet in Verbindung mit Fahrplänen, Öffnungszeiten usw.	Jim's train **arrives** at 4.30 p.m. Our office **opens** at 9 o'clock tomorrow morning.

Comparison of adjectives (Steigerung des Adjektivs)

Bildung	
Einsilbige und zweisilbige Adjektive werden durch Anhängen von *-er* und *-est* (germanische Steigerung) gesteigert. Aus ***y*** wird ***i***. Ein einzelner Endkonsonant wird nach einfachem betonten Vokal verdoppelt.	fast – faster – fastest easy – easier – easiest hot – hotter – hottest
Für die zweisilbigen gibt es keine eindeutigen Regeln. Gewöhnlich werden zweisilbige Adjektive, die auf *-y*, *-le*, *-er* oder *-ow* enden germanisch gesteigert, aber dies ist keine bindende Regel.	clever – cleverer – cleverest noble – nobler – noblest narrow – narrower – narrowest
Ebenfalls germanisch gesteigert werden zweisilbige Adjektive, wenn sie ihre Betonung auf der zweiten Silbe haben. Auch diese Regel ist nicht bindend. Daneben gibt es aber z. B. auch folgende Formen: *more / the most clever* oder *more / the most polite.*	polite – politer – politest sincere – sincerer – sincerest
Andere zweisilbige Adjektive werden in der Regel romanisch gesteigert.	nervous – more nervous – most nervous splendid – more splendid – most splendid
Alle drei- und mehrsilbigen Adjektive steigert man mit ***more*** und ***the most*** (romanische Steigerung).	expensive – more expensive – the most expensive beautiful – more beautiful – the most beautiful
Es gibt auch eine Reihe von unregelmäßigen Adjektiven.	good – better – the best bad – worse – the worst much / many – more – the most little – less – the least old – elder – the eldest (bei Familienangehörigen)

Verwendung	
Gleichheiten mit *as* ... *as*	He is **as** skilful **as** I. (so geschickt / fachmännisch wie ich)
Komparative mit *than*	Mike is **less** helpful **than** Tim. (weniger ... als)
Je ... desto...	**The** more I work with this machine **the** better I like it.
Superlativ beim Vergleich von mehr als 2 Objekten	The new electric drill is **the most** expensive tool in the brochure.

Adverbs (Adverbien)

Das Umstandswort (Adverb) dient zur näheren Beschreibung eines Eigenschaftswortes (Adjektives) oder eines Zeitwortes (Verbs).

Bildung	
Von der Wortbildung her unterscheidet man zwei Arten von Adverbien:	
1. ursprüngliche und durch Zusammensetzung gebildete Adverbien	often, quite, here, soon always, already, yesterday
2. durch Ableitung gebildete Abverbien: Hier wird an das Adjektiv ein *-ly* angehängt.	clear**ly**, usual**ly**, quick**ly**, easi**ly**, bad**ly**
Manche Adjektive und Adverbien sind	
– gleichlautend oder	daily, weekly, monthly, yearly, early, fast
– unterschiedlich.	good – **well**
Die Steigerung des Adverbs ist der Steigerung des Adjektivs vergleichbar.	carefully – more carefully – most carefully fast – faster – the fastest
Unregelmäßige Formen der Steigerung	well – better – best badly – worse – worst
y wird zu *i*	happy + -ly = happily
stummes *e* fällt weg	true + -ly = truly
Verdoppelung des Endkonsonanten	beautiful + -ly = beautifully

Verwendung	
Hier wird das Verb näher beschrieben.	The new machine **works well**.
Hier wird das Adjektiv *clear* näher beschrieben.	The small printer makes **remarkably clear** printouts.
Hier wird ein anderes Adverb *(quickly)* näher beschrieben.	Our new machine works **extremely quickly**.

Questions (Fragen)

Bildung: Alternative questions (Entscheidungsfragen)	
Entscheidungsfragen, die mit ja oder nein beantwortet werden können, werden im Englischen immer mit einem Hilfsverb oder einer Form von *to do* eingeleitet.	**Do** you smoke? **Did** you see Mr Miller yesterday?
Erstes Hilfsverb steht vor dem Subjekt.	**Have** you met Mr Miller?

Verwendung: Alternative questions (Entscheidungsfragen)	
Does Sally speak French?	Yes, she **does.** / No, she **doesn't**.
Can these tools be used in milling as well as in turning?	Yes, they **can.** / No, they **can't**.
Is the lathe set properly for this job?	Yes, it **is.** / No, it **isn't**.
Did he meet the chief executive at the fair?	Yes, he **did.** / No, he **didn't**.
Will you finish this job by Friday?	Yes, I **will.** / No, I **won't**.

Bildung: Questions with interrogatives (Fragen mit Fragefürwörtern)	
Die Fragen werden mit dem Fragefürwort eingeleitet. Vor dem Subjekt steht die Form von *to do (do / does; did)*, nach dem Subjekt steht das Vollverb im Infinitiv.	**How** do you **go** to work? **What** does she **do**?
Wird nach dem Subjekt oder einem Teil des Subjekts gefragt *(who, what, which, whose)*, so wird die Form von 'to do' nicht verwendet.	**What** is his job?

Verwendung: Questions with interrogatives (Fragen mit Fragefürwörtern)	
Fragewort ist Subjekt.	Who operates this machine? Which tools do you need for this job?
Fragewort ist nicht Subjekt.	Who / Whom did he meet at the exhibition? Which tool have you chosen for this job?

Word order (Wortstellung)

Bildung	
Die Wortstellung folgt im englischen Aussagesatz der Regel S – P – O: *Subject [They] – Predicate [have repaired] – Object [the broken tools]*	They have repaired the broken tools.
Dies gilt auch im Nebensatz (im Deutschen steht ja im Nebensatz das Prädikat am Ende).	I was told that they had repaired the broken tools.
Zeitbestimmungen stehen in der Regel entweder am Anfang oder am Ende des Satzes.	Yesterday they repaired the broken tools. They repaired the broken tools a week ago.
Mehrere Umstandsbestimmungen folgen gewöhnlich folgender Reihenfolge: *Manner – Place – Time* (Art und Weise – Ort – Zeit)	The mechanic repaired the tools in the shop yesterday. The mechanic repaired the tools quite skilfully in the shop last night.
Adverbien der Häufigkeit *(frequency)* können auch vor dem Hauptverb stehen: *always, ever, never, often, regularly, generally, rarely, seldom, occasionally.* Diese Adverbien stehen aber nach dem Verb *to be.* Bei Hilfsverben stehen die Adverbien der Häufigkeit zwischen dem Hilfsverb und dem Vollverb, bzw. nach dem Hilfsverb.	The service people **always** come on time. The service people **never** come without calling ahead. The service people **usually** check the machine in May. They are **generally** very reliable. The machine has **sometimes** had problems. With old machines you can **never** tell.

Defining relative clauses (Notwendige Relativsätze)

Ein Relativsatz ist notwendig, wenn die darin gegebene Information nötig ist, um deutlich zu machen, welche Person oder Sache gemeint ist.

Verwendung	
bei Personen ***who*** oder ***that***	The man **who / that** is operating the CNC machine usually works at the training centre.
bei Sachen ***which*** oder ***that***	The machine element **which / that** is moving the tool is the main spindle.
im Genitiv wird **whose** verwendet	Tom is an engineer **whose** skills are highly praised.
Das Relativpronomen als Satzobjekt kann weggelassen werden.	They called the technician (**who**) they saw standing at the milling machine. Sally read the operation plan (**which**) she had filled in this morning.

Non-defining relative clauses (Nicht notwendige Relativsätze)

Ein Relativsatz ist nicht notwendig, wenn die darin gegebene Information nicht nötig ist, um deutlich zu machen, welche Person oder Sache gemeint ist.

Verwendung	
Ein Relativsatz ist nicht notwendig, wenn die darin enthaltene Information nicht nötig ist, um zu erkennen, welche Person oder Sache gemeint ist.	**Sally's friend**, who is a skilled worker, **is doing his final exams this summer.** **The machine**, which is DEG's latest model, **works perfectly.**
Relativpronomen als Satzobjekt	**The three German apprentices**, who Sally met at the training centre, **are in their last year of training.**

Reported speech (Indirekte Rede)

Die indirekte Rede wird benutzt, wenn berichtet wird oder wurde, was jemand sagt oder sagte, bzw. macht oder machte (**Aussage**), wenn jemandem gesagt wird oder wurde, was geschehen soll (**Aufforderung**), oder wenn gefragt wird oder wurde, was geschieht oder geschah (**Frage**).

Bildung / Verwendung	
Steht das Wort, welches die indirekte Rede einleitet, in der Gegenwart, so ändern sich die Zeitformen der direkten Rede nicht.	"We **are measuring** this machined component." → Sally explains that they are measuring the machined component.
Steht das Wort, welches die indirekte Rede einleitet, in der Vergangenheit, so ändern sich die Zeitformen der direkten Rede wie folgt.	
Simple past und *present perfect* werden zu *past perfect*.	Sally **explained that they were measuring** the machined component. "The trainee **forgot** to set the revolution speed." → The master **noticed that the trainee had forgotten** to check the revolution speed.
Will wird zu *would*.	"The drilling operation **will** have to be prepared by marking out and centring." → Oliver **remembered** that the drilling operation **would** have to be prepared by marking out and centring.
Pronomen werden angepasst.	"**I can't** find the measuring equipment." → **Sally told** me that she **couldn't find** the measuring equipment.
Orts- und Zeitbestimmungen werden angepasst.	"**We** are seeing the other students **tomorrow**." → Oliver **added** that **they** were seeing the other students **the next day**. "I programmed the CNC machine **two days ago**." → He said that he had programmed the CNC machine **two days earlier**.
Aufforderungen werden in der indirekten Rede im Infinitiv wiedergegeben.	"**Use** HSS twist drills to drill metal stock." → The instructor **advised** Sally **to use** HSS twist drills to drill metal stock.
Bei Fragen wird die Wortstellung wie im Aussagesatz verwendet, d. h. am Anfang steht das Fragewort, dann folgen Subjekt, Prädikat und Objekt.	"Why has Oliver used the impact drilling function?" → The master **wanted to know why Oliver had used** the impact drilling function. "When did you sharpen the drill?" → He asked **when he had sharpened** the drill.
Die Umschreibung mit 'to do' wird nur bei Verneinungen angewendet.	"Why **don't** you call the salesman?" → She **asked** Sally why she **didn't call** the salesman.

Passive voice (Passiv)

Bildung	
Form von *to be* + *ed*-Endung (wie bei *simple past*) bei regelmäßigen Verben Form von *to be* + 3. Form bei unregelmäßigen Verben (*1. take / 2. took / 3.* ***taken***. Siehe Liste der unregelmäßigen Verben.)	The machine is equip**ped** with an emergency stop. He **is taken** to hospital.

Aktiv:	We *Subjekt*	recharge *Prädikat*	the batteries. *Objekt*
Passiv:	*Subjekt* The batteries	*Prädikat* are recharged	*Objekt (by-agent)* (by us).

Formen	
Present Tense (Simple)	The goods are transported every day.
Present Tense (Progressive)	The goods are being transported at the moment.
Past Tense (Simple)	The goods were transported yesterday.
Past Tense (Progressive)	The goods were being transported when the truck crashed.
Future 1 (will)	The goods will be transported next Monday, I promise.
Future 1 (going to)	The goods are going to be transported next month.
Present Perfect	The goods have already been transported.
Past Perfect	The goods had been transported before the order arrived.

Verwendung	
Das Passiv wird hauptsächlich verwendet, wenn der Ausführende (*by-agent*) unwichtig oder unbekannt ist oder nicht genannt werden soll. Deshalb wird der *by-agent* nur bei besonderer Betonung genannt.	The batteries **are recharged** (**by us**) when they are empty. A few years ago the first cars **were equipped** with airbags (**by the manufacturers**).
Die Passivkonstruktion wird besonders bei technischen Beschreibungen, Anweisungen in Handbüchern usw. verwendet. Bei der Umwandlung von Aktiv in Passiv wird das Objekt des Aktivsatzes zum Subjekt des Passivsatzes.	Aktiv: We recharge **the batteries**. Passiv: **The batteries** are recharged (by us).
Im Gegensatz zum Deutschen können bestimmte Verben im Englischen im Passiv verwendet werden. Die Übersetzung ins Deutsche erfolgt durch eine Konstruktion mit 'man'.	The fuel cell is considered to be a very promising invention. *(Man hält die Brennstoffzelle für eine sehr zukunftsweisende Erfindung.)*
Die Verlaufsform ist im Passiv nur im *Present Tense* und *Past Tense* gebräuchlich.	He **is being** taken to hospital. *(Er wird gerade ins Krankenhaus gebracht.)* He **was being** taken to hospital. *(Er wurde gerade ins Krankenhaus gebracht.)*

Gerund (Gerundium)

Wie im Deutschen kann man auch im Englischen aus einem Verb ein Substantiv machen. Beispiel: Durch häufiges **Messen** können Fehler vermieden werden.

Bildung		
Anhängen von *-ing* an den Infinitiv (Grundform) des Verbs	check + ing = checking drill + ing = drilling	
Verdoppelung eines Endkonsonanten (Mitlaut) nach einem Vokal (Selbstlaut)	put – putting control – controlling	travel – travelling refuel – refuelling
Stummes *e* fällt weg: aber: *agree – agreeing; see – seeing*	file – filing measure – measuring	

Verwendung	
Ein *gerund* kann Subjekt (Satzgegenstand) sein. Ebenso kann es die Funktion eines Objekts (Satzergänzung) haben.	**Filing** is rarely done by hand. He began **cleaning** the machine.
Nach Präpositionen (Verhältniswörtern) oder präpositionalen Ausdrücken steht ebenfalls das *gerund*.	He was afraid of **setting** the lever to a higher speed. They objected **to trying** the new software **without having** made a backup of the hard drive. The manager apologised **for being** late. They prevented the visitors **from entering** the production area without a hard hat. He was tired **of hearing** the same complaint.
Nach bestimmten Verben (Zeitwörtern) steht das *gerund*.	Try to **avoid overheating** the machine. The trainer **recommended turning** the machine off. We cannot **risk losing** this customer. Phillip **suggested using** a different production method.
Weitere Verben, die ein *gerund* nach sich haben.	enjoy, like, dislike, remember, stop He enjoys working in Italy.

Prepositions (Verhältniswörter)

Präpositionen geben das Verhältnis von Personen oder Dingen zueinander an (z. B. räumlich oder zeitlich).

Räumliche Präpositionen			
above	oberhalb	**into**	in … hinein
across	durch, quer durch	**off**	von … herunter, von … weg
at	neben, bei	**on, upon**	auf
opposite	gegenüber		

Räumliche Präpositionen			
behind	hinter	**out, out of**	aus, heraus
below	unter, unterhalb	**through**	durch
beneath	unter, unterhalb	**to**	zu, nach, in Richtung auf
beside	neben	**towards**	auf … zu (Richtung)
between	zwischen	**under**	unter
in	in	**within**	innerhalb

Zeitliche Präpositionen			
at	an, in	**from**	von
before	vor	**in**	am, im (Jahre)
between	zwischen	**on**	an, am
by	bis, bis spätestens	**since**	seit (Zeitpunkt)
during	während	**until, till**	bis
for	seit (Zeitraum)		

Modale Präpositionen und Präpositionen des Grundes			
according to	entsprechend, gemäß	**despite, in spite of**	trotz
because of	wegen	**with**	mit
besides	außer	**without**	ohne

Verwendung	
Laut Gebrauchsanweisung sollte **das Auswechseln** der Werkzeuge kein Problem darstellen.	**According to** the manual, there should be no problem **with** changing the tools.
… **hinten** im Raum … **unten** auf jeder Seite … **am** Eingang	The machine is **at** the back of the room. You'll find the number of the module **at** the bottom of every page. They were waiting **at** the entrance when it happened.
… ist **gegenüber / auf der anderen Seite** … rannte quer **über** die Straße	The printer is **across** the hall. The dog ran **across** the road.
… **zwischen** zwei Aktenmappen (räumlich) … **zwischen** 9 und 10 Uhr (zeitlich)	He found the letter **between** two folders. Why don't we meet **between** nine and ten?
… nur **während**	Smoking is allowed **during** the lunch break only!
… **von** … **bis**	He worked in Cardiff **from** December **until** January.
… **auf** der Straße	You should not park your car **in** the street.
… **bis** (spätestens)	You must be back **by** six o'clock at the latest.

Modal auxiliaries (Modale Hilfsverben)

Die modalen Hilfsverben werden im Englischen oft auch *defective* (d.h. unvollständige) *auxiliaries* genannt, denn sie können in der Regel nur das Präsens bilden. Für die übrigen Zeiten müssen Ersatzformen verwendet werden.

Bildung

Hilfsverb	**Ersatzform**	**deutsche Bedeutung**
can	to be able to	können, fähig sein, in der Lage sein
can	to be allowed to	dürfen, die Erlaubnis haben
may	to be allowed to	dürfen, die Erlaubnis haben
must	to have to	müssen
mustn't	not to be allowed to	nicht dürfen
needn't	not to have to	nicht brauchen, nicht müssen

Verwendung

Present Tense: You **must** read the installation instructions very carefully. After the installation is completed you **needn't reboot** your system.	**Simple past:** Yesterday he **had to** read the installation instructions twice because they were badly translated. I **didn't have to** reboot my system when I installed the screensaver the other day.
Present Tense: He **can** handle software problems quite well.	**Present perfect:** Up to now he **has been able** to handle software problems quite well.
Present Tense: You **must not smoke** in the office.	**Future:** You **will not be allowed** to smoke in the office.

Participles (Partizipien)

Bildung

Present Participle Infinitiv + *-ing*	Wind turbines are a **promising** option.
Past Participle 3. Stammform: *damaged* (*ed*-Endung wie bei *simple past*); (3. Form bei unregelmäßigen Verben: *1. know/2. knew/**3. known***. Siehe Liste der unregelmäßigen Verben.)	After the storm we had to replace three **damaged** solar panels.

Verwendung	
als Adjektiv	It is a **known** fact that solar power can be used as an alternative energy.
Partizipialkonstruktionen dienen zur **Verkürzung** von:	
1. adverbialen Nebensätzen mit Konjunktion *(when, although, as, while,…)*	When **planning** to build a home, you should consider using solar energy to heat your water.
2. adverbialen Nebensätzen ohne Konjunktion	**Unlocking** the warehouse door, he smelled gas.
3. Relativsätzen	Wind farms **built** along the coast are most efficient.
Partizipialkonstruktionen zur **Verknüpfung** von zwei Hauptsätzen.	The guide led us through the factory. He warned us to stay away from the machines. The guide led us through the factory **warning** us to stay away from the machines.

If-clauses (Bedingungssätze)

Man unterscheidet drei Arten von Bedingungen:
1. erfüllbare Bedingungen
2. Bedingungen, deren Erfüllung unwahrscheinlich ist
3. Bedingungen, die überhaupt nicht mehr erfüllt werden können.

Zusätzlich zu den drei Typen gibt es noch einen weiteren Typ, der Regeln und Naturgesetze ausdrückt.

Bildung und Verwendung	
Type I (probable condition)	If you **lubricate** this machine regularly, it **will run** smoothly for a long time. *(Wenn man die Maschine regelmäßig ölt, läuft sie lange Zeit problemlos.)*
Type II (improbable condition)	If you **lubricated** this machine regularly, **it would** run smoothly for a long time. *(Wenn man diese Maschine regelmäßig ölen würde, würde sie für lange Zeit problemlos laufen.)*
Type III (impossible condition)	If you **had lubricated** this machine regularly, it **would have run** smoothly for a long time. *(Wenn man die Maschine regelmäßig geölt hätte, wäre sie lange Zeit problemlos gelaufen (jetzt läuft sie aber offensichtlich nicht mehr).*
Rules and laws of nature Hier handelt es sich im eigentlichen Sinne nicht mehr um eine Bedingung, sondern um ein Naturgesetz. Deshalb steht im Hauptsatz auch kein Futur, sondern ebenfalls das Präsens.	If you **heat** ice, it **melts**. *(Wenn man Eis erhitzt, schmilzt es.)*
Besonderheit bei Formen von *to be*	If I **were** you, I wouldn't use a combination spanner here.

Chronological word list

Videotraining: Englische Aussprache

Perfekte englische Aussprache leicht gemacht: Mit dem Lernprogramm zur englischen Lautschrift können Sie alle Laute einüben. Wählen Sie einfach in der Navigation rechts den entsprechenden Reiter (*Vowels* oder *Consonants*) aus und klicken Sie dann auf das gewünschte phonetische Symbol. Sprechen Sie die Wörter laut nach.

Unter www.klett.de geben Sie bitte den Code unter der Abbildung rechts ein.

n633sn

Abkürzungen und Zeichen

etw. = etwas
Pl. = Plural
jmdn., jmdm. = jemanden, jemandem
sb. = somebody
sth. = something
BE = britisches Englisch
AE = amerikanisches Englisch
◎ = Vokabeln zu den Hörtexten

Module 1 Introduction to vehicles

vehicle [ˈvɪəkl] Fahrzeug
to own [əʊn] besitzen
manufacturer [ˌmanjəˈfæktʃrə] Hersteller
size [saɪz] Größe
engine [ˈendʒɪn] Motor
environmentalist [ɪnˌvaɪrnˈmentlɪst] Umweltschützer(in)
to argue [ˈɑːgjuː] diskutieren, streiten, argumentieren
environment [ɪnˈvaɪərnmənt] Umwelt
confidence [ˈkɒnfɪdns] Selbstvertrauen, Sicherheit
classification [ˌklæsɪfɪˈkeɪʃn] Gruppierung, Klassifizierung
brand [brænd] Marke
acceleration [əkˌseləˈreɪʃn] Beschleunigung

A | Classifying cars

common [ˈkɒmən] gebräuchlich, häufig
performance [pəˈfɔːməns] Leistung
to depend on [dɪˈpend‿ɒn] abhängig sein von
purpose [ˈpɜːpəs] Zweck, Absicht
microcar [ˈmaɪkrəʊˌkɑː] Kleinstwagen
compact executive [ˌkɒmpækt‿ɪgˈzekjətɪv] Kompaktwagen
compact MPV car [ˌkɒmpækt‿empiːˈviː ˌkɑː] Mehrzweckfahrzeug
executive car [ɪgˈzekjətɪv ˌkɑː] Vorstandswagen
SUV (sport utility vehicle) [ˌesjuːˈviː, ˌspɔːt juːˈtɪləti ˌvɪəkl] Geländewagen
maintenance costs [ˈmeɪntnəns ˌkɒsts] Wartungskosten
space [speɪs] Raum
luggage [ˈlʌgɪdʒ] Gepäck
comfortable [ˈkʌmftəbl] bequem
cruise control [ˈkruːz kənˌtrəʊl] Tempomat, Geschwindigkeitsregelanlage
in-car computer [ˌɪnkɑː kəmˈpjuːtə] Bordcomputer
saloon car [səˈluːn ˌkɑː] Limousine
latest [ˈleɪtɪst] neueste(r, s)
feature [ˈfiːtʃə] Ausstattung, Einrichtung; Eigenschaft, Merkmal, Kennzeichen
weight [weɪt] Gewicht
versatility [ˌvɜːsəˈtɪləti] Vielseitigkeit, Flexibilität

four-wheel drive [ˌfɔ:wi:l 'draɪv] Allradantrieb
business ['bɪznɪs] Unternehmen, Geschäft
abbreviation [əˌbri:vi'eɪʃn] Abkürzung
to mention ['menʃn] erwähnen, nennen

radio host ['reɪdiəʊ ˌhəʊst] Radiomoderator(in)
car dealer ['kɑ: ˌdi:lə] Autohändler(in)
particularly [pə'tɪkjələli] besonders
to mean [mi:n] bedeuten
running costs ['rʌnɪŋ ˌkɒsts] Unterhaltskosten, laufende Kosten
boot (BE) / trunk (AE) [bu:t, trʌŋk] Kofferraum
though [ðəʊ] jedoch, obwohl
to favour ['feɪvə] bevorzugen
chiefly ['tʃi:fli] hauptsächlich, vornehmlich
amazing [ə'meɪzɪŋ] erstaunlich, toll, beeindruckend
to share [ʃeə] teilen

target ['tɑ:gɪt] Ziel
to climb [klaɪm] klettern

B | Parts of a vehicle

bonnet (BE) / hood (AE) ['bɒnɪt, hʊd] Motorhaube
bumper ['bʌmpə] Stoßstange
headlight ['hedlaɪt] Scheinwerfer
windscreen (BE) / windshield (AE) ['wɪnskri:n, 'wɪnʃi:ld] Windschutzscheibe
radiator grille (BE) / grill (AE) ['reɪdieɪtə ˌgrɪl, grɪl] Kühlergrill
wing (BE) / fender (AE) [ˌ'wɪŋ, f'ʌndər] Kotflügel
door handle ['dɔ: ˌhændl] Türgriff
wiper ['waɪpə] Scheibenwischer
tyre (BE) / tire (AE) [taɪə] Reifen
wing mirror (BE) / side mirror (AE) ['wɪŋ ˌmɪrə, 'saɪd ˌmɪrə] Außenspiegel
accelerator (BE) / gas pedal [ək'seləreɪtə, 'gæs ˌpedl] Gas(pedal)
brake [breɪk] Bremse
clutch [klʌtʃ] Kupplung
dashboard ['dæʃbɔ:d] Armaturenbrett
glove compartment ['glʌv kəmˌpɑ:tmənt] Handschuhfach
gear stick ['gɪə ˌstɪk] Schalthebel, Schaltknüppel
speedometer [spi:'dɒmətə] Tachometer, Geschwindigkeitsmesser
rod [rɒd] Stange
gear [gɪə] Gang

C | Choosing the right car

choice [tʃɔɪs] Wahl, Auswahl
wide [waɪd] breit, weit
range [reɪndʒ] Spektrum, Palette, Angebot
to afford [ə'fɔ:d] sich leisten
to consider [kən'sɪdə] betrachten, in Betracht ziehen
issue ['ɪʃu:] Frage, Thema, Problem
to run [rʌn] *(hier:)* unterhalten; laufen, rennen
fuel consumption ['fju:əl kənˌsʌmʃn] Kraftstoffverbrauch
economical [ˌi:kə'nɒmɪkl] kostengünstig, wirtschaftlich
average ['ævrɪdʒ] durchschnittlich
majority [mə'dʒɒrəti] Mehrzahl, Mehrheit
tax [tæks] Steuer
cubic capacity [ˌkju:bɪk kə'pæsəti] Hubraum
to emit [ɪ'mɪt] ausstoßen, emittieren
plenty ['plenti] viel
air conditioning ['eə kənˌdɪʃnɪŋ] Klimaanlage
rough [rʌf] grob, rau
terrain [tə'reɪn] Gelände, Terrain
suspension [sə'spenʃn] Federung, Aufhängung
waste [weɪst] Abfall, Verschwendung
general ['dʒenrl] allgemein
sense [sens] Sinn, Bedeutung
to accommodate [ə'kɒmədeɪt] unterbringen, Platz bieten, berücksichtigen
to be worth [bi 'wɜ:ð] wert sein, lohnen
rear seat [rɪə 'si:t] Rücksitz
to fold [fəʊld] falten
suitable ['su:təbl] geeignet, passend
automatic transmission [ˌɔ:təmætɪk trænz'mɪʃn] Automatikgetriebe
to tow [təʊ] ziehen, schleppen, abschleppen
trailer ['treɪlə] Anhänger
to exceed [ɪk'si:d] übersteigen, überschreiten
manual transmission [ˌmænjuəl trænz'mɪʃn] Schaltgetriebe

Module A Meeting people

confident ['kɒnfɪdənt] sicher, selbstbewusst
appropriate [ə'prəʊpriət] angemessen, passend

A | Saying "hello" and "goodbye" and giving your name

existing [ɪg'zɪstɪŋ] bestehend
trade fair ['treɪd ˌfeə] Messe
job fair ['dʒɒb ˌfeə] Jobmesse

B | Intercultural awareness

politeness [pə'laɪtnəs] Höflichkeit
pleasantry ['plezntri] höfliche Floskel
profitability [ˌprɒfɪtəbɪləti] Rentabilität
to chat [tʃæt] sich unterhalten
health [helθ] Gesundheit
pleasant ['pleznt] angenehm
to waste time [ˌweɪst 'taɪm] Zeit verschwenden
impression [ɪm'preʃn] Eindruck
conference ['kɒnfrəns] Tagung
minutes ['mɪnɪts] Protokoll
polite [pə'laɪt] höflich
usual ['ju:ʒl] gewöhnlich, üblich, normal
upset [ʌp'set] verärgert
stranger ['streɪndʒə] Unbekannte(r)
on the other hand [ɒn ði‿'ʌðə ˌhænd] andererseits
in charge of [ɪn 'tʃɑ:dʒ‿əv] verantwortlich für
cash register ['kæʃ ˌredʒɪstə] Kasse
given name [ˌgɪvn 'neɪm] Vorname
family name [ˌfæmli 'neɪm] Nachname
to intend [ɪn'tend] beabsichtigen

Module 2 Tools in a car workshop

tool [tu:l] Werkzeug, Gerät
workshop ['wɜ:kʃɒp] Werkstatt
variety [və'raɪəti] Vielfalt, Auswahl
screwdriver ['skru:ˌdraɪvə] Schraubendreher
spanner [ˌspænə] Schraubenschlüssel
torque wrench [ˌtɔ:k 'renʃ] Drehmomentschlüssel
adjustable wrench [əˌdʒʌstəbl 'renʃ] Rollgabelschlüssel
colleague ['kɒli:g] Kollege, Kollegin
customer ['kʌstəmə] Kunde, Kundin
tyre pressure gauge ['taɪə ˌpreʃə ˌgeɪdʒ] Reifendruckmesser
hacksaw ['hæksɔ:] Bügelsäge
pliers [plaɪəz] Zange
lug wrench [lʌg 'renʃ] Radmutternschlüssel
wrench ['renʃ] Gabelschlüssel

A | Tools – Usage and safety

ring spanner ['rɪŋ ˌspænə] Ringschlüssel
extension [ɪk'stenʃn] Verlängerung
combination pliers [ˌkɒmbɪneɪʃn 'plaɪəz] Kombizange
chisel ['tʃɪzl] Meißel
Allen key ['ælən ˌki:] Innensechskantschlüssel, Inbusschlüssel
ratchet ['rætʃɪt] Ratsche
punch [pʌnʃ] Körner
synonym ['sɪnənɪm] bedeutungsgleiches Wort, Synonym
hexagon socket key [ˌheksəgɒn 'sɒkɪt ˌki:] Innensechskantschlüssel, Inbusschlüssel
loop [lu:p] Öse, Schlinge
device [dɪ'vaɪs] Gerät, Vorrichtung
handle ['hændl] Griff
socket wrench ['sɒkɪt ˌrenʃ] Steckschlüssel
to loosen ['lu:sn] lösen, lockern
to tighten ['taɪtn] festziehen, spannen
nut [nʌt] Schraubenmutter
bolt [bəʊlt] (Schrauben-)Bolzen
to shape [ʃeɪp] formen, prägen
purpose ['pɜ:pəs] Zweck, Absicht
to drill [drɪl] bohren
file [faɪl] Feile
grinder ['graɪndə] Schleifmaschine
Phillips screwdriver [ˌfɪlɪps 'skru:ˌdraɪvə] Kreuzschlitzschraubendreher
slot screwdriver [ˌslɒt 'skru:ˌdraɪvə] Schlitzschraubendreher
tyre lever ['taɪə ˌli:və] Reifenmontierhebel
to jumble ['dʒʌmbl] durcheinander werfen / bringen
familiar [fə'mɪliə] vertraut
tangle ['tæŋgl] Durcheinander, Knäuel
ground [graʊnd] Boden, Fußboden
power cord ['paʊə ˌkɔ:d] Stromkabel
to unplug [ʌn'plʌg] ausstecken
to insulate ['ɪnsjəleɪt] isolieren, dämmen
glasses ['glɑ:sɪz] Brille
extinguisher [ɪk'stɪŋgwɪʃə] Feuerlöscher
first aid [ˌfɜ:st‿'eɪd] Erste Hilfe
wheel nut ['wi:l ˌnʌt] Radmutter
impact driver ['ɪmpækt ˌdraɪvə] Schlagbohrer
jack [dʒæk] Wagenheber
to jack up [ˌdʒæk‿'ʌp] aufbocken, hochheben
lifting platform ['lɪftɪŋ ˌplætfɔ:m] Hebebühne
torque [tɒrk] Drehmoment

pattern ['pætn] Muster, Schema
wear [weər] Abrieb, Verschleiß
spare tyre [ˌspeə 'taɪə] Ersatzreifen
wheel hub ['wi:l ˌhʌb] Radnabe
sketch [sketʃ] Skizze, kurze Darstellung
drawer [drɔ:] Schublade
necessary ['nesəsri] nötig, notwendig

B | Types of tools

power drill ['paʊə ˌdrɪl] elektrische Bohrmaschine
column drill ['kɒləm ˌdrɪl] Standbohrmaschine
wire stripper ['waɪə ˌstrɪpə] Abisolierzange
centre lathe [ˌsentə 'leɪð] Drehmaschine
angle grinder ['æŋgl ˌgraɪndə] Winkelschleifer, Flex
milling machine ['mɪlɪŋ məʃi:n] Fräsmaschine
power tool ['paʊə ˌtu:l] Elektrowerkzeug
prompt [prɒmt] Hinweis, Stichwort
to recommend [ˌrekə'mend] empfehlen
body ['bɒdi] Karosserie; Körper
dent [dent] Beule, Delle
oil filter wrench [ˌɔɪlˌfɪltə 'renʃ] Ölfilterschlüssel
funnel ['fʌnl] Trichter

C | Hand tools and power tools

precision [prɪ'sɪʒn] Genauigkeit
fan belt ['fæn ˌbelt] Keilriemen
for instance [fər‿'ɪnstns] zum Beispiel
compressed air [kəmˌprest‿'eə] Druckluft
skill [skɪl] Fähigkeit, Fertigkeit
to require [rɪ'kwaɪə] verlangen, benötigen
metal bar ['metl ˌbɑ:] Metallwerkstück
drive mechanism ['draɪv ˌmekənɪzm] Antriebsmechanismus
gear [gɪə] Gang; Zahnrad
to fail [feɪl] scheitern, versagen
maintenance ['meɪntnəns] Wartung
to sharpen ['ʃɑ:pn] schleifen, schärfen, anspitzen
to increase [ɪn'kri:s] (an)steigen, steigern, erhöhen
injury ['ɪndʒri] Verletzung

◎
assembly [ə'sembli] Montage
engineering [ˌendʒɪ'nɪərɪŋ] Technik, Maschinenbau
recent ['ri:snt] kürzlich, neulich
to diversify [daɪ'vɜ:sɪfaɪ] aufteilen, verändern
leisure ['leʒə] Freizeit, freie Zeit
renowned [rɪ'naʊnd] angesehen, berühmt
award [ə'wɔ:d] Auszeichnung, Preis
to register ['redʒɪstə] eintragen, registrieren

Module B Your company

regardless of [rɪ'gɑ:dləs‿əv] ungeachtet, unabhängig von
furniture ['fɜ:nɪtʃə] Möbel
likely ['laɪkli] wahrscheinlich
supplier [sə'plaɪə] Lieferant, Zulieferer
skill [skɪl] Fähigkeit, Fertigkeit
to express [ɪk'spres] zum Ausdruck bringen
grease [gri:s] Fett
ratchet ['rætʃɪt] Knarre
tool [tu:l] Werkzeug

A | The company and its departments

department [dɪ'pɑ:tmənt] Abteilung
to distribute [dɪ'strɪbju:t] vertreiben, verteilen
damaged ['dæmɪdʒd] beschädigt
employee [ɪm'plɔɪi:] Beschäftigte(r), Angestellte(r)
accounts [ə'kaʊnts] Buchhaltung
raw materials [ˌrɔ: mə'tɪəriəlz] Rohstoffe
human resources [ˌhju:mən 'ri:zɔ:sɪz] Personalabteilung
to recruit [rɪ'kru:t] rekrutieren, anstellen
purchasing (department) ['pɜ:tʃəsɪŋ (dɪˌpɑ:tmənt)] Einkauf(sabteilung)
to maintain [meɪn'teɪn] warten, in Stand halten
R&D (Research and Development) [rɪˌsɜ:tʃ‿ən dɪ'veləpmənt] Forschungs- und Entwicklungsabteilung
to develop [dɪ'veləp] entwickeln
in charge of [ɪn 'tʃɑ:dʒ‿əv] zuständig für

◎
to cover ['kʌvə] abdecken, bedecken
headquarters [ˌhed'kwɔ:təz] Hauptsitz
to found [faʊnd] gründen
family-owned [ˌfæmli'əʊnd] in Familienbesitz
manufacturing plant [ˌmænjə'fæktʃərɪŋ ˌplɑ:nt] Produktionsbetrieb
currently ['kʌrəntli] derzeit, zurzeit, momentan
braking system ['breɪkɪŋ ˌsɪstəm] Bremsanlage
internship (AE) ['ɪntɜ:rnʃɪp] Praktikum
device [dɪ'vaɪs] Gerät
cylinder head ['sɪlɪndə ˌhed] Zylinderkopf
to inspect [ɪn'spekt] prüfen
operating system ['ɒpreɪtɪŋ ˌsɪstəm] Betriebssystem
interface ['ɪntəfeɪs] Schnittstelle, Interface

B | Tasks and responsibilities

task [tɑ:sk] Aufgabe
responsibility [rɪˌspɒnsəˈbɪləti] Verantwortung
to manufacture [ˌmænjəˈfæktʃə] herstellen
to occur [əˈkɜ:] vorkommen
apprenticeship [əˈprentɪsʃɪp] Ausbildung, Lehre
branch [brɑ:nʃ] Niederlassung
precision [prɪˈsɪʒn] Genauigkeit
essential [ɪˈsenʃl] wesentlich, unbedingt erforderlich
to drill [drɪl] bohren
to shape [ʃeɪp] formen, fräsen
to polish [ˈpɒlɪʃ] abschleifen, polieren
to intervene [ˌɪntəˈvi:n] einschreiten
to vibrate [vaɪˈbreɪt] vibrieren
workpiece [ˈwɜ:kpi:s] Werkstück
to monitor [ˈmɒnɪtə] überwachen, kontrollieren, beobachten
to measure [ˈmeʒə] messen
dimension [daɪˈmenʃn] Maß, Abmessung
lubricated [ˈlu:brɪkeɪtɪd] geölt
lubricating oil [ˈlu:brɪkeɪtɪŋ‿ɔɪl] Schmieröl
generally [ˈdʒenrəli] normalerweise, im Allgemeinen
shift [ʃɪft] Schicht
to prefer [prɪˈfɜ:] bevorzugen
rarely [ˈreəli] selten
accuracy [ˈækjərəsi] Genauigkeit
to include [ɪnˈklu:d] einbeziehen, einschließen
to swap [swɒp] tauschen

C | Organisational structure

to employ [ɪmˈplɔɪ] beschäftigen
to support [səˈpɔ:t] unterstützen
to split [splɪt] aufteilen
core area [ˌkɔ: ˈeəriə] Kernbereich
to report to [rɪˈpɔ:t tu:] unterstehen
to belong to [bɪˈlɒŋ tu:] gehören zu
experienced [ɪkˈspɪəriənst] erfahren
capable [ˈkeɪpəbl] fähig
closely [ˈkləʊsli] eng

Module 3 The body shop

body shop [ˈbɒdi ˌʃɒp] Karosseriewerkstatt, Rohbau
stamping shop [ˈstæmpɪŋ ˌʃɒp] Presswerk
to join [dʒɔɪn] zusammenfügen
car body [ˈkɑ: ˌbɒdi] Karosserie
to surround [səˈraʊnd] umgeben
to protect [prəˈtekt] (be)schützen
body shell [ˈbɒdi ˌʃel] Rohbaukarosserie
passenger safety cell [ˌpæsɪndʒə ˈseɪfti ˌsel] Rohbauzelle
B-pillar [ˈbi: ˌpɪlə] B-Säule
boot [bu:t] Kofferraum
floor pan [ˈflɔ: ˌpæn] Bodenwanne, Bodenblech
front wing [ˌfrʌnt ˈwɪŋ] vorderer Kotflügel
sill [sɪl] Schweller
wheel arch [ˈwi:l‿ɑ:tʃ] Radlauf

A | What happens in the body shop?

statement [ˈsteɪtmənt] Aussage(satz)
approximately [əˈprɒksɪmətli] ungefähr, etwa
to weld [weld] schweißen
joining technology [ˈdʒɔɪnɪŋ tekˌnɒlədʒi] Fügetechnik
upper [ˈʌpə] Ober-, obere
bonnet [ˈbɒnɪt] Motorhaube
boot lid [ˌbu:t ˈlɪd] Kofferraumdeckel
hang-on part [ˌhæŋ‿ˈɒn ˌpɑ:rt] Aufhängeteil
assembly shop [əˈsembli ˌʃɒp] Montagehalle
body in white [ˌbɒdi‿ɪn ˈwaɪt] Rohkarosserie, Karosseriegerippe
to stamp [stæmp] stanzen, pressen
sheet metal [ˌʃi:t ˈmetl] Blech
raw [rɔ:] roh
metal shell [ˈmetl ˌʃel] Metallgehäuse
saloon [səˈlu:n] Limousine
sub-assembly [ˈsʌbəˌsembli] Baugruppe, Vormontage
front unit [ˌfrʌnt ˈju:nɪt] Vorbau
rear unit [ˌrɪə ˈju:nɪt] Heckteil
centre unit [ˌsentə ˈju:nɪt] Bodenteil
side panel [ˌsaɪd ˈpænl] Seitenwand
jig pallet [ˈdʒɪg ˌpælɪt] Bauteilträgerwagen
conveyor [kənˈveɪə] Förderband, Förderanlage
frame [freɪm] Rahmen
to ensure [ɪnˈʃɔ:] gewährleisten, sicherstellen
joint [dʒɔɪnt] Verbindung, Fuge
weld [weld] Schweißnaht
to check [tʃek] (über)prüfen
tank lid [ˌtæŋk ˈlɪd] Tankverschluss
to screw [skru:] schrauben
geometry welding station [dʒiˌɒmɪtri ˈweldɪŋ ˌsteɪʃn] Geometrie-Schweißstation
underbody [ˈʌndəˌbɒdi] Unterbau
smoothness [ˈsmu:θnəs] Glätte
measurement [ˈmeʒəmənt] Abmessung, Maß
deviation [ˌdi:viˈeɪʃn] Abweichung
according to [əˈkɔ:dɪŋ tə] entsprechend, gemäß, laut

B | Joining technologies in the body shop

adhesive bonding [əd,hi:sɪv 'bɒndɪŋ] Klebeverbindung
brazing ['breɪzɪŋ] Hartlöten
rivet joining ['rɪvɪt ,dʒɔɪnɪŋ] Vernieten
spot welding ['spɒt ,weldɪŋ] Punktschweißen
laser welding ['leɪzə ,weldɪŋ] Laserschweißen
filler metal ['fɪlə ,metl] Schweißzusatzwerkstoff
fusion process ['fju:ʒn ,prəʊses] Fusionsprozess
edge [edʒ] Kante, Rand
thickness ['θɪknəs] Stärke, Dicke
to clamp together [,klæmp tə'geðə] zusammenklemmen
current ['kʌrnt] Strom
glue [glu:] Klebstoff
tailgate ['teɪlgeɪt] Hecktür, Heckklappe
to assure [ə'ʃʊə] zusichern, garantieren

to melt [melt] schmelzen
to fuse [fju:z] verschmelzen
rivet gun ['rɪvɪt ,gʌn] Nietpistole
self-piercing rivet [,self,pɪəsɪŋ 'rɪvɪt] selbststanzende Niete
to crimp [krɪmp] quetschen, kräuseln
stiff [stɪf] steif, fest
cold-formed [,kəʊld'fɔ:md] kaltumformt
hot-formed [,hɒt'fɔ:md] warmumformt
high-strength [,haɪ'streŋθ] hochfest, hochwiderstandsfähig
challenging ['tʃælɪndʒɪŋ] herausfordernd
thermoplastic [,θɜ:mə'plæstɪk] Thermoplast
fibre reinforced [,faɪbə,ri:ɪn'fɔ:st] faserverstärkt
to cherry-pick [tʃeri,pɪk] sich das Beste herauspicken

C | Materials to build the body in white

longitudinal member [,lɒndʒɪ'tju:dɪnl ,membə] Längsträger
toxic fume [,tɒksɪk 'fju:mz] giftiges Gas
to inhale [ɪn'heɪl] einatmen

Module C Telephoning

line [laɪn] *(hier:)* Branche
to fix [fɪks] reparieren
leak [li:k] undichte Stelle
to install [ɪn'stɔ:l] installieren, montieren, anschließen
complaint [kəm'pleɪnt] Beschwerde, Reklamation
order ['ɔ:də] Bestellung, Auftrag
necessary ['nesəsri] nötig, notwendig
delay [dɪ'leɪ] Verspätung, Verzögerung
breakdown ['breɪkdaʊn] Panne, Störung
solution [sə'lu:ʃn] Lösung
confused [kən'fju:zd] verwirrt
crash [kræʃ] Absturz

A | Making and receiving a phone call

meeting ['mi:tɪŋ] Sitzung, Besprechung
country code ['kʌntri ,kəʊd] Ländervorwahl
area code ['eəriə ,kəʊd] Vorwahl
extension number [ɪk'stenʃn ,nʌmbə] Durchwahl
suitable ['su:təbl] geeignet, passend
polite [pə'laɪt] höflich
although [ɔ:l'ðəʊ] obwohl
to remain [rɪ'meɪn] bleiben
request [rɪ'kwest] Bitte
grateful ['greɪtfl] dankbar
whereas [weə'ræz] während
impolite [,ɪmpə'laɪt] unhöflich

B | Giving information over the phone

valuable ['væljuəbl] *(hier:)* wichtig
figures ['fɪgəz] Nummern, Daten
clearly ['klɪəli] deutlich, klar

confused [kən'fju:zd] verwirrt
to divert [daɪ'vɜ:rt] weiterleiten
extractor fan [ɪk'stræktə fæn] Dunstabzug
hob [hɒb] Kochfeld
diameter [daɪ'æmətə] Durchmesser
socket ['sɒkɪt] Steckdose

Module 4 The paint shop

paint shop ['peɪnt ,ʃɒp] Lackiererei
plant [plɑ:nt] Fabrik(anlage)
body ['bɒdi] *(hier:)* Karosserie
coat [kəʊt] Beschichtung, Überzug
body shop ['bɒdi ,ʃɒp] Karosseriewerkstatt, Rohbau
to dip [dɪp] eintauchen
body shell ['bɒdi ,ʃel] Rohbaukarosserie
base coat ['beɪs ,kəʊt] Basislack
clear coat ['klɪə ,kəʊt] Klarlack
undercoat ['ʌndəkəʊt] Grundierung

A | Stages in painting a car

rust [rʌst] Rost
UV rays [ˌju:'vi: ˌreɪz] UV-Strahlen
grease [gri:s] Fett
spotless ['spɒtləs] fleckenfrei
initial [ɪ'nɪʃl] erste(r, s)
conveyor [kən'veɪə] Fließband
application [ˌæplɪ'keɪʃn] *(hier:)* Anstrich
layer [leɪə] Schicht
to immerse [ɪ'mɜ:s] eintauchen
cavity ['kævəti] Hohlraum
to seal [si:l] abdichten, versiegeln
curing ['kjʊərɪŋ] Aushärten
seam [si:m] Saum
corrosion resistant [kəˌrəʊʒn rɪ'zɪstnt] korrosionsbeständig
to sand [sænd] absanden, abschmirgeln
primer ['praɪmə] Grundierung
to apply [ə'plaɪ] *(hier:)* auftragen
cleanliness ['klenlɪnəs] Sauberkeit
dust-free [ˌdʌst'fri:] staubfrei
treatment ['tri:tmənt] Aufbereitungsverfahren
underbody ['ʌndəˌbɒdi] Unterbau

B | A visit to the paint shop

to obtain [əb'teɪn] erhalten, erzielen

◎
impurity [ɪm'pjʊərəti] Verunreinigung
blemish ['blemɪʃ] Makel
airlock chamber ['eəlɒk ˌtʃeɪmbə] Luftschleuse
ostrich ['ɒstrɪtʃ] (Vogel-)Strauß
solvent ['sɒlvənt] Lösungsmittel
to inject [ɪn'dʒekt] einspritzen
nozzle ['nɒzl] Spritzdüse
assembly [ə'sembli] Montage
to register ['redʒɪstə] eintragen, registrieren

C | Quality problems in the paint shop

blister ['blɪstə] (Guss-)Blase
chip [tʃɪp] Abplatzung
crater ['kreɪtə] Krater (Autolack)
dent [dent] Delle, Beule
drip [drɪp] Tropfen
orange peel ['ɒrɪndʒ ˌpi:l] Orangenschale
scratch [skrætʃ] Kratzer
droplet ['drɒplət] Tröpfchen
pin hole ['pɪn ˌhəʊl] Nadelloch

Module D Written communication

essential [ɪ'senʃl] sehr wichtig
workshop ['wɜ:kʃɒp] Werkstatt
efficiently [ɪ'fɪʃntli] effizient
necessary ['nesəsri] nötig, notwendig
enquiry [ɪn'kwaɪəri] Anfrage
promptly ['prɒmtli] sofort
offer ['ɒfə] Angebot
order ['ɔ:də] Bestellung
to replace [rɪ'pleɪs] ersetzen
lesser ['lesə] geringer
advantage [əd'vɑ:ntɪdʒ] Vorteil
disadvantage [ˌdɪsəd'vɑ:ntɪdʒ] Nachteil
confidential [ˌkɒnfɪ'denʃl] vertraulich
time-consuming ['taɪmkənˌsju:mɪŋ] zeitraubend
reliable [rɪ'laɪəbl] zuverlässig
misunderstanding [ˌmɪsʌndə'stændɪŋ] Missverständnis

A | Enquiries

enquiry [ɪn'kwaɪəri] Anfrage
car mechanic ['kɑ: mɪˌkænɪk] Automechaniker(in)
appropriate [ə'prəʊpriət] angemessen, passend
salutation [ˌsælјə'teɪʃn] *(Brief:)* Anrede
supplier [sə'plaɪə] Lieferant, Zulieferer
subject line ['sʌbdʒɪkt ˌlaɪn] *(Brief:)* Betreffzeile
bonnet ['bɒnɪt] Motorhaube
family-run [ˌfæmli'rʌn] in Familienbesitz, familiengeführt
to expand [ɪk'spænd] erweitern, vergrößern
catalogue ['kætəlɒg] Katalog
discount ['dɪskaʊnt] Rabatt, Preisnachlass
terms of payment [ˌtɜ:mz‿əv 'peɪmənt] Zahlungsbedingungen
delivery times [dɪ'lɪvri ˌtaɪmz] Lieferzeiten
sample ['sæmpl] Muster
closing phrase ['kləʊzɪŋ ˌfreɪz] Schlussworte
complimentary close [ˌkɒmplɪmentri 'kləʊz] Grußformel (am Briefende)
to recommend [ˌrekə'mend] empfehlen
cost estimate ['kɒst‿ˌestɪmət] Kostenvoranschlag
to configure [kən'fɪgə] konfigurieren
to appreciate [ə'pri:ʃieɪt] etw. (wert)schätzen
supply from stock [səˌplaɪ frəm 'stɒk] ab Lager liefern

B | Offers

bank transfer ['bæŋk ˌtrænsfɜː] Banküberweisung
working day [ˌwɜːkɪŋ 'deɪ] Werktag
attachment [ə'tætʃmənt] Anhang
stock [stɒk] Bestand, Vorrat
to require [rɪ'kwaɪə] benötigen
ratchet ['rætʃɪt] Knarre, Ratsche
socket ['sɒkɪt] Stecknuss
(adjustable) wrench [(əˌdʒʌstəbl) 'renʃ] (verstellbarer) Schraubenschlüssel
screwdriver ['skruːˌdraɪvə] Schraubendreher
to dispatch [dɪ'spætʃ] versenden, verschicken
DIY (Do-it-yourself) [ˌdiːaɪ'waɪ] Heimwerken

C | Orders

sum [sʌm] Betrag
consignment [kən'saɪnmənt] Lieferung, Sendung
torque wrench ['tɔːk ˌrenʃ] Drehmomentschlüssel

Module 5 Chassis and drivetrain

chassis ['ʃæsi] Fahrgestell
drivetrain ['draɪvtreɪn] Antrieb
underneath [ˌʌndə'niːθ] unter
alloy ['ælɔɪ] Legierung
body ['bɒdi] *(hier:)* Karosserie
suspension [sə'spenʃn] Federung, Aufhängung
unit-body construction [ˌjuːnɪtˌbɒdi kən'strʌkʃn] Schalenbauweise, Monocoque
powertrain ['paʊətreɪn] Antrieb
transmission [trænz'mɪʃn] Getriebe
drive axle ['draɪv ˌæksl] Antriebsachse
RWD (rear wheel drive) [ˌɑːdʌbljuː'diː, ˌrɪəˌwiːl 'draɪv] Hinterradantrieb
FWD (front wheel drive) [ˌefdʌbljuː'diː, ˌfrʌntˌwiːl 'draɪv] Vorderradantrieb

A | Chassis construction

fuel [fjuːəl] Kraftstoff, Treibstoff
coach [kəʊtʃ] Kutsche
to steer [stɪə] lenken, steuern
smooth [smuːð] glatt, reibungslos
bump [bʌmp] *(hier:)* Unebenheit
shock absorber ['ʃɒk‿əbˌzɔːbə] Stoßdämpfer
confusing [kən'fjuːzɪŋ] verwirrend, irreführend
SUV (sport utility vehicle) [ˌesjuː'viː, ˌspɔːt juː'tɪləti ˌvɪəkl] Geländewagen

B | Suspension systems

to dive [daɪv] (vorne) absinken; tauchen
to accelerate [ək'seləreɪt] beschleunigen
roadholding ['rəʊdˌhəʊldɪŋ] Straßenlage
cornering ['kɔːnrɪŋ] Kurvenfahren
to grip [grɪp] greifen, haften
bend [bend] Kurve
to brake [breɪk] bremsen
to dampen ['dæmpən] dämpfen
to squat [skwɒt] (hinten) absenken; (hin)hocken
to bounce [baʊns] federn, hüpfen
to compress [kəm'pres] zusammendrücken
to extend [ɪk'stend] (sich) ausdehnen
strut [strʌt] Federbein
spring [sprɪŋ] Feder
coil spring ['kɔɪl ˌsprɪŋ] Spiralfeder, Sprungfeder
leaf spring ['liːf ˌsprɪŋ] Blattfeder
torsion bar spring ['tɔːʃn ˌbɑː ˌsprɪŋ] Drehstabfeder
air spring ['eə ˌsprɪŋ] Luftfeder
to disperse [dɪ'spɜːs] auflösen, zerstreuen
rebound ['riːbaʊnd] Rückfederung, Rückprall

C | Drivetrain

torque [tɔːk] Drehmoment
driveshaft ['draɪv ˌʃɑːft] Antriebswelle
clutch [klʌtʃ] Kupplung
RPM (revs / revolutions per minute) [ˌɑːpiː'em, ˌrevz / revəˌluːʃnz pə 'mɪnɪt] Umdrehungen pro Minute
to stall [stɔːl] Motor abwürgen
gear stick ['gɪə ˌstɪk] Schalthebel

Module E Applications

to sign [saɪn] unterschreiben
application [ˌæplɪ'keɪʃn] Bewerbung
opportunity [ˌɒpə'tjuːnəti] Möglichkeit
to decide [dɪ'saɪd] sich entscheiden
branch [brɑːnʃ] Niederlassung
advert ['ædvɜːt] Anzeige
letter of application [ˌletər‿əv‿ˌæplɪ'keɪʃn] Bewerbungsschreiben
CV (curriculum vitae) [ˌsiː'viː, kəˌrɪkjələm 'viːtaɪ] Lebenslauf
job interview ['dʒɒb‿ˌɪntəvjuː] Vorstellungsgespräch
employment contract [ɪm'plɔɪmənt ˌkɒntrækt] Arbeitsvertrag
to invite [ɪn'vaɪt] einladen

skill [skɪl] Fähigkeit
industry ['ɪndəstri] *(hier:)* Branche
hard-working [ˌhɑ:d'wɜ:kɪŋ] fleißig

A | Job advertisements

recently ['ri:sntli] kürzlich
willing to learn [ˌwɪlɪŋ tə 'lɜ:n] lernbereit, lernwillig
responsible [rɪ'spɒnsəbl] verantwortlich
punctual ['pʌŋktʃuəl] pünktlich
pleasant ['pleznt] angenehm, sympathisch
working atmosphere [ˌwɜ:kɪŋ‿'ætməsfɪə] Arbeitsklima
valid ['vælɪd] gültig
troubleshooting ['trʌblˌʃu:tɪŋ] Fehlersuche
to maintain [meɪn'teɪn] bewahren
leading ['li:dɪŋ] führend
supplier [sə'plaɪə] Lieferant, Zulieferer
vacancy ['veɪknsi] freie Stelle
fault [fɒlt] Fehler
assembly [ə'sembli] Montage
summary ['sʌmri] Zusammenfassung

B | Letter of application

letter of application [ˌletər‿əv‿ˌæplɪ'keɪʃn] Bewerbungsschreiben
internship (AE) ['ɪntɜ:rnʃɪp] Praktikum
theoretical [θɪə'retɪkl] theoretisch
to gain [geɪn] *(hier:)* sammeln
reliable [rɪ'laɪəbl] zuverlässig
to consider [kən'sɪdə] betrachten
to require [rɪ'kwaɪə] benötigen

C | The CV

advice [əd'vaɪs] Rat
tip [tɪp] Tipp, Hinweis
to highlight ['haɪlaɪt] betonen
strength [streŋθ] Stärke
achievement [ə'tʃi:vmənt] Leistung
supervisor ['su:pəvaɪzə] Leiter(in), Kontrolleur(in)
truth [tru:θ] Wahrheit
embarrassing [ɪm'bærəsɪŋ] peinlich
to spell [spel] buchstabieren
grateful ['greɪtfl] dankbar

Module 6 Assembly and logistics

assembly [ə'sembli] Montage, Herstellung
logistics [lə'dʒɪstɪks] Logistik, Versorgung
body ['bɒdi] *(hier:)* Karosserie
pre-defined [ˌpri:dɪ'faɪnd] im Voraus festgelegt
smooth [smu:ð] glatt, reibungslos

A | The assembly process

to adjust [ə'dʒʌst] einstellen, anpassen
to assemble [ə'sembl] montieren, zusammenbauen
to insulate ['ɪnsjəleɪt] isolieren
to mount [maʊnt] montieren, einbauen
conveyor system [kən'veɪə ˌsɪstəm] Förderanlage
beam [bi:m] (Licht-)Strahl
steering column ['stɪərɪŋ ˌkɒləm] Lenksäule

chassis number ['ʃæsi ˌnʌmbə] Fahrgestellnummer
door trim line ['dɔ: ˌtrɪm ˌlaɪn] Türverkleidungs-fließband
wiring harness ['waɪərɪŋ] Kabelbaum
headliner ['hedˌlaɪnə] Dachverkleidung
pre-assembly [ˌpri:ə'sembli] Vormontage
manually ['mænjuəli] von Hand, manuell
to seal [si:l] abdichten, versiegeln
windshield ['wɪnʃi:ld] Windschutzscheibe
rear window [ˌrɪə 'wɪndəʊ] Heckscheibe
side panel ['saɪd ˌpænl] Seitenwand
edge protection ['edʒ prəˌtekʃn] Kantenschutz
piping ['paɪpɪŋ] Verrohrung
to hoist ['hɔɪst] hochheben
fuel gauge unit ['fju:əl ˌgeɪdʒ ˌju:nɪt] Kraftstoffanzeigeeinheit
powertrain ['paʊətreɪn] Antrieb(sstrang)
to raise [reɪz] anheben
rubber sealant [ˌrʌbə 'si:lənt] Kautschuk-dichtungsmasse
power steering [ˌpaʊə 'stɪərɪŋ] Servolenkung
windscreen wiper tank ['wɪnskri:n ˌwaɪpə ˌtæŋk] Scheibenwischertank
to align [ə'laɪn] fluchten, ausrichten
rolls test ['rəʊlz ˌtest] Abrollprüfung

B | The logistics department

aisle [aɪl] Fahrstraße, Gang
dock [dɒk] Rampe
dolly ['dɒli] Rollbrett
stock [stɒk] Lagerware

supplier [sə'plaɪə] Lieferant
warehouse ['weəhaʊs] Lagerhalle
collapsible [kə'læpsəbl] zusammenklappbar
forklift (truck) [ˌfɔ:klɪft ('trʌk)] Gabelstapler
high-bay storage [ˌhaɪbeɪ 'stɔ:rɪdʒ] Hochregallager

C | Problems and solutions in logistics

delay [dɪ'leɪ] Verspätung, Verzögerung
inventory [ɪn'vəntri] (Lager-)Bestand
congestion [kən'dʒestʃn] Stau

congested [kən'dʒestɪd] verstopft, überfüllt
aisle [aɪl] Gang
rack [ræk] Regal, Ständer
facility [fə'sɪləti] Einrichtung, Anlage
grommet ['grɒmɪt] Halterungsöse
washer ['wɒʃə] Dichtung(sring)

Module F Socialising

relationship [rɪ'leɪʃnʃɪp] Beziehung

A | Small talk

subsidiary [səb'sɪdɪəri] Tochtergesellschaft
trainee [ˌtreɪ'ni:] Auszubildende(r)
appropriate [ə'prəʊpriət] angemessen, passend

B | Small talk in business

to dismiss [dɪ'smɪs] nicht beachten
common ground [ˌkɒmən 'graʊnd] gemeinsame Basis
trivial ['trɪviəl] unwichtig
uneasiness [ʌn'i:zinəs] Unbehagen
tension ['tenʃn] Spannung
to avoid [ə'vɔɪd] vermeiden
awkward ['ɔ:kwəd] peinlich, unangenehm

C | Eating out

well-done [ˌwel'dʌn] (Steak) gut durchgebraten
pork [pɔ:k] Schweinefleisch
side order ['saɪd‿ɔ:də] Beilage
bill [bɪl] Rechnung
game [geɪm] *(hier:)* Wild
dessert [dɪ'zɜ:t] Nachtisch, Nachspeise
fork [fɔ:k] Gabel
main dish [ˌmeɪn 'dɪʃ] Hauptgericht
beef [bi:f] Rindfleisch
poultry ['pəʊltri] Geflügel
starter ['stɑ:tə] Vorspeise
tip [tɪp] Trinkgeld

Module 7 Engine basics

combustion engine [kəm'bʌstʃn‿ˌendʒɪn] Verbrennungsmotor
camshaft ['kæmʃɑ:ft] Nockenwelle
connecting rod (conrod) ['kɒnektɪŋ rɒd ('kɒnrɒd)] Pleuelstange
crankshaft ['kræŋkʃɑ:ft] Kurbelwelle
flywheel ['flaɪwi:l] Schwungrad
to ignite [ɪg'naɪt] zünden
sump [sʌmp] Ölwanne
valve [vælv] Ventil
shaft [ʃɑ:ft] Welle
cam ['kæm] Nocken
piston ['pɪstn] Kolben
to ignite [ɪg'naɪt] zünden
lubrication [ˌlu:brɪ'keɪʃn] Schmierung
revolving [rɪ'vɒlvɪŋ] sich drehend

A | How a basic combustion engine works

fuel [fju:əl] Brennstoff, Treibstoff
intake valve ['ɪnteɪk ˌvælv] Einlassventil
exhaust valve [ɪg'zɔ:st ˌvælv] Auslassventil
alloy ['ælɔɪ] Legierung
sliding seal [ˌslaɪdɪŋ 'si:l] Gleitdichtung
combustion chamber [kəm'bʌstʃn ˌtʃeɪmbə] Brennkammer, Verbrennungsraum
induction stroke [ɪn'dʌkʃn ˌstrəʊk] Ansaugtakt
compression stroke [kəm'preʃn ˌstrəʊk] Druckhub
dead centre [ˌded 'sentə] genaue Mitte
spark plug ['spɑ:k ˌplʌg] Zündkerze
to expand [ɪk'spænd] sich ausdehnen
exhaust [ɪg'zɔ:st] Abgase
four-stroke cycle [ˌfɔ:strəʊk 'saɪkl] Viertakt-Zyklus
intake stroke ['ɪnteɪk ˌstrəʊk] Ansaughub

horizontally opposed [hɒrɪˌzɒntli‿ə'pəʊzd] Boxeranordnung; horizontal gegenüberliegend

B | The diesel engine

refined [rɪ'faɪnd] raffiniert
glow plug ['gləʊ ˌplʌg] Glühkerze
to evaporate [ɪ'væpreɪt] verdunsten

mileage ['maɪlɪdʒ] Reichweite, Benzinverbrauch

to shift [ʃɪft] schalten
acceleration [əkˌselə'reɪʃn] Beschleunigung
brake horsepower (bhp) [ˌbreɪk 'hɔ:sˌpaʊə, ˌbi:eɪtʃ'pi:] Pferdestärke (PS, am Rad gemessen)
cubic capacity [ˌkju:bɪk kə'pæsəti] Hubraum
displacement [dɪ'spleɪsmənt] Hubraum
manual transmission [ˌmænjuəl trænz'mɪʃn] Schaltgetriebe
performance [pə'fɔ:məns] Leistung
rated output [ˌreɪtɪd'aʊtpʊt] Nennleistung
torque [tɔ:k] Drehmoment
turbocharged ['tɜ:bəʊˌtʃɑ:dʒd] turboaufgeladen
gear stick ['gɪə ˌstɪk] Schalthebel

C | Turbo chargers

turbo charger ['tɜ:bəʊ ˌtʃɑ:dʒə] Turbolader
to spin [spɪn] (sich) drehen
to melt [melt] schmelzen
delay [dɪ'leɪ] Verzögerung, Verspätung
manufacturer [ˌmænjə'fæktʃrə] Hersteller
to launch [lɔ:nʃ] einführen (auf dem Markt)
cylinder bore ['sɪlɪndə ˌbɔ:] Zylinderbohrung
particulate filter [pə'tɪkjələt ˌfɪltə] Partikelfilter
catalytic converter [ˌkætəlɪtɪk kən'vɜ:tə] Katalysator

Module G Presentations

required [rɪ'kwaɪəd] erforderlich, verlangt, erwünscht
service ['sɜ:vɪs] Dienstleistung
positional sketch [pəˌzɪʃnl 'sketʃ] Technologie-schema
to develop [dɪ'veləp] entwickeln
to improve [ɪm'pru:v] besser werden, sich verbessern
meeting ['mi:tɪŋ] Sitzung, Besprechung
graph [grɑ:f] Diagramm
chart [tʃɑ:t] Schaubild, Diagramm
preparation [ˌprepr'eɪʃn] Vorbereitung
to explain [ɪk'spleɪn] erklären
sales pitch ['seɪlz ˌpɪtʃ] Verkaufspräsentation
reason ['ri:zn] Grund

A | Preparing a presentation

audience ['ɔ:diəns] Zuhörer *(Pl.)*, Publikum
body language ['bɒdi ˌlæŋgwɪdʒ] Körpersprache
handout ['hændaʊt] Informationsblatt
to prepare [prɪ'peə] vorbereiten
visual aids [ˌvɪʒuəl'eɪdz] Anschauungsmaterial, visuelle Hilfen
to describe [dɪ'skraɪb] beschreiben
summary ['sʌmri] Zusammenfassung
overview ['əʊvəvju:] Überblick
to reinforce [ˌri:ɪn'fɔ:s] verstärken, betonen
guidelines ['gaɪdlaɪnz] Leitlinien

B | Describing materials and products

sturdy ['stɜ:di] robust
shiny ['ʃaɪni] glänzend
robust [rə'bʌst] stabil
state-of-the-art [ˌsteɪtəv ði'ɑ:t] auf dem neuesten Stand (der Technik)
user-friendly [ju:zə'frendli] benutzerfreundlich
reliable [rɪ'laɪəbl] zuverlässig
hand-crafted [ˌhæn'krɑ:ftɪd] handgefertigt

C | Delivering a presentation

carpenter ['kɑ:pɪntə] Tischler(in), Zimmerer, Zimmerin
furniture ['fɜ:nɪtʃə] Möbel
trade show ['treɪd ʃəʊ] Messe
lounger ['laʊndʒə] Liegesessel
softwood ['sɒftˌwʊd] Weichholz
timber ['tɪmbə] Holz, Bauholz
joiner ['dʒɔɪnə] Tischler(in), Schreiner(in)
convenience [kənˌvi:niəns] Zweckmäßigkeit, Bequemlichkeit
varnish ['vɑ:nɪʃ] Lack
durable ['djʊərəbl] haltbar
polyurethane coating [ˌpɒlɪjʊərəθeɪn 'kəʊtɪŋ] Polyurethanbeschichtung

Module 8 Alternative engines

fuel ['fju:əl] Brennstoff, Treibstoff
manufacturer [ˌmænjə'fæktʃrə] Hersteller
combustion engine [kəm'bʌstʃnˌendʒɪn] Verbrennungsmotor
research [rɪ'sɜ:tʃ] Forschung
development [dɪ'veləpmənt] Entwicklung
fuel cell ['fju:əl ˌsel] Brennstoffzelle
hydrogen ['haɪdrədʒən] Wasserstoff
to propel [prə'pel] antreiben
fuel tank ['fju:əl ˌtæŋk] Treibstofftank
filling station ['fɪlɪŋ ˌsteɪʃn] Tankstelle
to emit [ɪ'mɪt] ausstoßen

distance ['dɪstns] Entfernung
harm [hɑ:m] Schaden
poisonous ['pɔɪznəs] giftig
emission [ɪ'mɪʃn] Emission, Ausstoß
environment [ɪn'vaɪrnmənt] Umwelt; Umgebung

A | The electric vehicle

to recharge [ˌri:'tʃɑ:dʒ] wieder aufladen
exhaust [ɪg'zɔ:st] Auspuff
rechargeable [ˌri:'tʃɑ:dʒəbl] wieder aufladbar
controller [kən'trəʊlə] Steuergerät, Regler
carbon dioxide [ˌkɑ:bn daɪ'ɒksaɪd] Kohlendioxid
carbon monoxide [ˌkɑ:bn mə'nɒksaɪd] Kohlenmonoxid
nitrogen ['naɪtrədʒən] Stickstoff
sulphur ['sʌlfə] Schwefel
hydrocarbon [ˌhaɪdrə'kɑ:bn] Kohlenwasserstoff
exhaust [ɪg'zɔ:st] Auspuff
kit [kɪt] Ausrüstung, Set

B | The fuel cell car

fuel cell stack [ˌfju:əl sel 'stæk] Brennstoffzellenstapel

compressed [kəm'prest] komprimiert
regenerative braking [rɪˌdʒenrətɪv 'breɪkɪŋ] Bremsenergierückgewinnung
to run out of sth. [ˌrʌn‿'aʊt‿əv ˌsʌmθɪŋ] ausgehen, zu Ende gehen

apprentice [ə'prentɪs] Auszubildende(r), Lehrling
accelerator [ək'seləreɪtə] Gaspedal
torque [tɔ:k] Drehmoment
speeding ticket ['spi:dɪŋ ˌtɪkɪt] Strafzettel für zu schnelles Fahren

C | The hybrid car

propulsion system [prə'pʌlʃn ˌsɪstəm] Antriebssystem
to plug in [ˌplʌg‿'ɪn] einstecken, an Strom anschließen
fuel consumption ['fju:əl kənˌsʌmʃn] Kraftstoffverbrauch

to refuel [ˌri:'fju:əl] nachtanken
air pollution ['eə pəˌlu:ʃn] Luftverschmutzung
pedestrian [pə'destriən] Fußgänger(in)
flammable ['flæməbl] leicht entzündlich

to discontinue [ˌdɪskən'tɪnju:] einstellen, beenden, auslaufen lassen

Module H Dealing with customers

to deal with ['di:l wɪð] umgehen mit, sich befassen mit
customer ['kʌstəmə] Kunde, Kundin
success [sək'ses] Erfolg
appointment [ə'pɔɪntmənt] Termin
enquiry [ɪn'kwaɪəri] Anfrage
complaint [kəm'pleɪnt] Beschwerde
essential [ɪ'senʃl] sehr wichtig
builder ['bɪldə] Bauunternehmer(in)
building site ['bɪldɪŋ ˌsaɪt] Baustelle
to explain [ɪk'spleɪn] erklären

A | Complaints

solution [sə'lu:ʃn] Lösung
delay [dɪ'leɪ] Verzögerung, Verspätung
price reduction ['praɪs rɪˌdʌkʃn] Preisnachlass
faulty ['fɒlti] fehlerhaft, defekt, mangelhaft
damaged ['dæmɪdʒd] beschädigt
unsatisfactory [ʌnˌsætɪs'fæktri] unbefriedigend, ungenügend, unzureichend
to replace [rɪ'pleɪs] ersetzen
air freight ['eə ˌfreɪt] Luftfracht
surplus ['sɜ:pləs] überschüssig
credit note ['kredɪt ˌnəʊt] Gutschrift
specific [spə'sɪfɪk] bestimmt, spezifisch

access ['ækses] Zugang
to log on [ˌlɒg‿'ɒn] sich anmelden, sich einwählen
migration [maɪ'greɪʃn] *(hier:)* Umstellung
to redo [ˌri:'du:] noch einmal machen
to look into sth. [ˌlʊk‿'ɪntə] einer Sache nachgehen

technician [tek'nɪʃn] Techniker(in)
plumbing and heating [ˌplʌmɪŋ‿ənd 'hi:tɪŋ] Heizung und Sanitär

B | Customer service

upset [ʌp'set] verärgert
to avoid [ə'vɔɪd] vermeiden
argument ['ɑ:gjəmənt] Streit, Auseinandersetzung
to calm sb. [kɑ:m] jmdn. beruhigen
to reassure [ˌri:ə'ʃʊə] beruhigen

patient ['peɪʃnt] geduldig
to solve [sɒlv] lösen
frame of mind [ˌfreɪm‿əv 'maɪnd] geistige Verfassung
furious ['fjʊəriəs] wütend
to affect [ə'fekt] Einfluss nehmen auf
mood [mu:d] Laune
to blame sb. / sth. for sth. [bleɪm] jmdm. / etw. die Schuld an etw. geben, jmdn. verantwortlich machen
excuse [ɪk'skju:s] Ausrede

Module 9 Safety

demand [dɪ'mɑ:nd] Nachfrage
fatality [fə'tæləti] Todesopfer
term [tɜ:m] Begriff
safety features ['seɪfti ˌfi:tʃəz] Sicherheitsausstattung
to refer (to) [rɪ'fɜ:] sich beziehen (auf)
occupant ['ɒkjəpənt] (Fahrzeug-)Insasse
anti-lock braking system [ˌæntɪlɒk 'breɪkɪŋ ˌsɪstəm] Antiblockiersystem (ABS)
deformation zone [ˌdi:fɔ:'meɪʃn ˌzəʊn] Knautschzone
intelligent speed adaptation [ɪnˌtelɪdʒnt ˌspi:d ˌædæp'teɪʃn] intelligente Geschwindigkeitsanpassung
brake assist system [ˌbreɪk‿ə'sɪst ˌsɪstəm] Bremsassistenzsystem
traction control system [ˌtrækʃn kən'trəʊl ˌsɪstəm] Antriebsschlupfregelung
seat belt ['si:t ˌbelt] Sicherheitsgurt
collision warning / avoidance [kə'lɪʒn ˌwɔ:nɪŋ / ə̮ˌvɔɪdns] Zusammenstoßwarnung / -vermeidung
autonomous cruise control system [ɔ:ˌtɒnəməs ˌkru:z kən'trəʊl ˌsɪstəm] Tempomat

A | Safety in SUVs

SUV (sport utility vehicle) [ˌesju:'vi:, ˌspɔ:t ju:'tɪləti ˌvɪəkl] Geländewagen
estate car [ɪ'steɪt ˌkɑ:] Kombi
fuel-efficient [ˌfju:əl‿ɪ'fɪʃnt] kraftstoffsparend
elevated ['eləveɪtɪd] erhöht
tall [tɔ:l] hoch
centre of gravity [ˌsentər‿əv 'grævəti] Schwerpunkt
to tip over [ˌtɪp‿'əʊvə] umkippen
to swerve [swɜ:v] ausweichen

obstacle ['ɒbstəkl] Hindernis
to compensate ['kɒmpənseɪt] ausgleichen, kompensieren
to reinforce [ˌri:ɪn'fɔ:s] verstärken
ultra high-strength steel [ˌʌltrə ˌhaɪstreŋθ 'sti:l] ultrahochfester Stahl
frame [freɪm] Rahmen
side panel ['saɪd ˌpænl] Seitenverkleidung, Seitenteil
stiff [stɪf] starr, fest
gyro sensor ['dʒaɪrəʊ ˌsensə] Kreiselsensor
to monitor ['mɒnɪtə] überwachen, kontrollieren
anti-skid traction control [ˌæntɪskɪd 'trækʃn kənˌtrəʊl] Traktionskontrolle und Antischleudersystem
to steer [stɪə] lenken, steuern

seat belt tensioner ['si:tbelt ˌtenʃnə] Gurtstraffer
to tighten ['taɪtn] festziehen, spannen
to deploy [dɪ'plɔɪ] auslösen
inflatable [ɪn'fleɪtəbl] aufblasbar
side curtain ['saɪd ˌkɜ:tn] Seitenairbag
to eject [ɪ'dʒekt] herausschleudern
B-pillar ['bi: ˌpɪlə] B-Säule
to pierce [pɪəs] durchstechen
rescue ['reskju:] Rettung
vital ['vaɪtl] unerlässlich, lebensnotwendig
hazard warning light [ˌhæzəd 'wɔ:nɪŋ ˌlaɪt] Warnblinklicht

B | Safety features

headrest ['hedrest] Kopfstütze
blind-spot detection [ˌblaɪndspɒt dɪ'tekʃn] Totwinkelerkennung
lane-departure warning [ˌleɪn dɪ'pɑ:tʃə ˌwɔ:nɪŋ] Spurhalteassistent
seat belt force limiter [ˌsi:tbelt 'fɔ:s ˌlɪmɪtə] Gurtkraftbegrenzung
driver drowsiness sensor [ˌdraɪvə 'draʊzɪnəs ˌsensə] Fahrerzustandserkennung
seat-occupant sensing [ˌsi:tˌɒkjəpənt 'sensɪŋ] Sitzbelegungsfunktion
whiplash ['wɪplæʃ] Schleudertrauma
slack [slæk] Lockerung, Durchhang
to illuminate [ɪ'lu:mɪneɪt] ausleuchten, erleuchten
adjoining [ə'dʒɔɪnɪŋ] angrenzend
radiator grille [ˌreɪdieɪtə 'grɪl] Kühlergrill
inflator [ɪn'fleɪtə] (Airbag-)Gasgenerator
steering wheel ['stɪərɪŋ ˌwi:l] Lenkrad

to inflate [ɪn'fleɪt] aufblasen
spark [spɑ:k] Funken
to store [stɔ:] verstauen
sodium azide [ˌsəʊdiəm‿'æzaɪd] Natriumazid (NaN_3)
nitrogen ['naɪtrədʒən] Stickstoff
inflated [ɪn'fleɪtɪd] aufgeblasen

C | Testing safety features

pedestrian [pə'destriən] Fußgänger(in)

Module 10 Wheels and brakes

wheel ['wi:l] Rad, Felge
brake [breɪk] Bremse
motor vehicle mechatronics technician [ˌməʊtə vɪəkl ˌmekə'trɒnɪks tekˌnɪʃn] Kraftfahrzeug-mechatroniker(in)
service mechanic [sɜ:vɪs məˌkænɪk] Kfz-Service-mechaniker(in)
power unit ['paʊə ˌju:nɪt] Antriebseinheit
drivetrain ['draɪvtreɪn] Antrieb(sstrang)
suspension [sə'spenʃn] Federung, Aufhängung
tyre (BE) / tire (AE) [taɪə] Reifen
wear and tear ['weər ənd 'teər] Verschleiß
light alloy wheel [ˌlaɪt‿ælɔɪ 'wi:l] Leichtmetallrad, Alufelge
rim [rɪm] Felge, Radkranz

A | Wheels and tyres

wheel bolt ['wi:l ˌbəʊlt] Radschraube
chock [tʃɒk] Keil
tread [tred] (Reifen-)Profil
jack [dʒæk] Wagenheber
jacking point ['dʒækɪŋ ˌpɔɪnt] Wagenheber-Ansatzpunkt
tyre pressure ['taɪə ˌpreʃə] Reifendruck
to tighten ['taɪtn] anziehen, festziehen
crosswise ['krɒswaɪz] über Kreuz
tightening torque [ˌtaɪtnɪŋ 'tɔ:k] Anziehdrehmoment
to lock [lɒk] *(hier:)* einrasten

snow tyre [ˌsnəʊ 'taɪə] Winterreifen
all-season tyre [ˌɔ:l‿'sɪ:zən 'taɪə] Ganzjahresreifen
grip [grɪp] Haftung
rubber compound [ˌrʌbə 'kɒmpaʊnd] Gummi-mischung
braking distance ['breɪkɪŋ ˌdɪstns] Bremsweg

B | Disc brakes and drum brakes

disc brake ['dɪsk ˌbreɪk] Scheibenbremse
drum brake ['drʌm ˌbreɪk] Trommelbremse
brake caliper ['breɪk ˌkælɪpə] Bremssattel
brake booster ['breɪk ˌbu:stə] Bremskraftverstärker
brake shoe ['breɪk ˌʃu:] Bremsbacke, Bremsklotz
brake pedal ['breɪk ˌpedl] Bremspedal
lining ['laɪnɪŋ] *(hier:)* Bremsbelag
brake fluid ['breɪk ˌflu:ɪd] Bremsflüssigkeit
compressed [kəm'prest] komprimiert
piston ['pɪstn] Kolben
pad [pæd] Klotz, Belag
ventilation slot [ˌventɪ'leɪʃn ˌslɒt] (Be-)Lüftungsschlitz
dust cap ['dʌst ˌkæp] Staubschutz
wheel hub ['wi:l ˌhʌb] Radnabe
stud [stʌd] Spike
slide pin ['slaɪd ˌpɪn] Bolzenführung
brake disc ['breɪk ˌdɪsk] Bremsscheibe
return spring [rɪ'tɜ:n ˌsprɪŋ] Rückstellfeder

C | Braking systems

brake drum ['breɪk ˌdrʌm] Bremstrommel
wheel-brake cylinder ['wi:lbreɪk ˌsɪlɪndə] Radbremszylinder
brake pressure reducer ['breɪk ˌpreʃə rɪˌdju:sə] Bremsdruckminderer / -regler
brake circuit ['breɪk ˌsɜ:kɪt] Bremskreis
tandem master cylinder [ˌtændem 'mɑ:stə ˌsɪlɪndə] Tandemhauptzylinder
lug nut ['lʌg ˌnʌt] Radmutter
wheel wrench ['wi:l ˌrenʃ] Kreuzschlüssel
c-clamp ['si: ˌklæmp] Schraubzwinge
shock absorber ['ʃɒk‿əbˌzɔ:bə] Stoßdämpfer
bracket ['brækɪt] Klammer
to mount [maʊnt] montieren, befestigen
torque wrench ['tɔ:k ˌrenʃ] Drehmomentschlüssel

Module 11 Car systems

cooling system ['ku:lɪŋ ˌsɪstəm] Kühlsystem
exhaust system [ɪg'zɔ:st ˌsɪstəm] Abgassystem
overheating [ˌəʊvə'hi:tɪŋ] Überhitzung
liquid ['lɪkwɪd] Flüssigkeit
harmful ['hɑ:mfl] schädlich
environment [ɪn'vaɪərnmənt] Umwelt; Umgebung
vent [vent] Lüftungsöffnung

A | The liquid cooling system

antifreeze ['æntɪfri:z] Frostschutzmittel
fluid ['flu:ɪd] Flüssigkeit
coolant ['ku:lənt] Kühlmittel
fan [fæn] Ventilator, Gebläse, Lüfter
heater core ['hi:tər kɔ:r] Wärmetauscher
hose ['həʊz] Schlauch
pressure cap [ˌpreʃə 'kæp] Druckverschluss
radiator ['reɪdieɪtə] Kühler
radiator drain cock [ˌreɪdieɪtə 'dreɪn ˌkɒk] Kühlerablasshahn
spark plug ['spɑ:k ˌplʌg] Zündkerze
valve [vælv] Ventil
to ignite [ɪg'naɪt] zünden
tube [tju:b] Rohr
boiling point ['bɔɪlɪŋ ˌpɔɪnt] Siedepunkt
blade [bleɪd] Schraubenblatt, Flügel
lid [lɪd] Deckel
pressure ['preʃə] Druck

grille [ˌgrɪl] (Kühler-)Grill
to release [re'li:s] freisetzen, entlassen

B | The exhaust system

tail pipe ['teɪl ˌpaɪp] Endrohr
toxic waste [ˌtɒksɪk 'weɪst] Schadstoff(e)
combustion [kəm'bʌstʃn] Verbrennung
exhaust manifold [ɪg'zɔ:st ˌmænɪfəʊld] Abgaskrümmer
catalytic converter [ˌkætəlɪtɪk kən'vɜ:tə] Katalysator
silencer ['saɪlənsə] Schalldämpfer, Auspufftopf
oxygen sensor ['ɒksɪdʒən ˌsensə] Sauerstoffsensor, Lambdasonde
bracket ['brækɪt] Klammer
gasket ['gæskɪt] Dichtungsring
clamp [klæmp] Klammer
combustion chamber [kəm'bʌstʃn ˌtʃeɪmbə] Brennkammer, Verbrennungsraum
hydrocarbon [ˌhaɪdrə'kɑ:bn] Kohlenwasserstoff
nitrogen oxide [ˌnaɪtrədʒən‿'ɒksaɪd] Stickstoffoxid
emission [ɪ'mɪʃn] Emission
muffler ['mʌflə] Schalldämpfer
to emit [ɪ'mɪt] ausstoßen

C | Dealing with problems in the systems

work placement ['wɜ:k ˌpleɪsmənt] Praktikum
plug [plʌg] Stecker, Stöpsel
flush solution ['flʌʃ səˌlu:ʃn] Spüllösung
bucket ['bʌkɪt] Eimer
funnel ['fʌnl] Trichter
cleaning solution ['kli:nɪŋ səˌlu:ʃn] Reinigungslösung

to leak [lɪ:k] tropfen, lecken
crack [kræk] Riss

Module 12 Maintenance and repairs

maintenance ['meɪntnəns] Wartung
to have a car serviced [ˌhæv‿ə ˌkɑ: 'sɜ:vɪst] ein Auto warten lassen
warranty ['wɒrnti] Garantie, Gewährleistung
schedule ['ʃedju:l] Terminplan
manufacturer [ˌmænjə'fæktʃrə] Hersteller
service record ['sɜ:vɪs ˌrekɔ:d] Service-Scheckheft
glove compartment ['glʌv kəmˌpɑ:tmənt] Handschuhfach

A | Having your car serviced

breakdown ['breɪkdaʊn] Panne, Motorschaden
brake fluid ['breɪk ˌflu:ɪd] Bremsflüssigkeit
brake pad ['breɪk ˌpæd] Bremsklotz, Bremsbelag
wash / wipe system [ˌwɒʃ'waɪp ˌsɪstəm] Scheibenwasch- / wischanlage
windscreen ['wɪnskri:n] Windschutzscheibe
wiper blade ['waɪpə ˌbleɪd] Scheibenwischerblatt
nozzle ['nɒzl] Spritzdüse
joint [dʒɔɪnt] Verbindung
oxygen sensor ['ɒksɪdʒən ˌsensə] Sauerstoffsensor, Lambdasonde

B | Do it yourself

comprehensive [ˌkɒmprɪ'hensɪv] umfassend
scratch [skrætʃ] Kratzer
hood (AE) [hʊd] Motorhaube
fender (AE) [fʌndər] Kotflügel
dent [dent] Delle, Beule
panel ['pænl] Blech
windshield (AE) ['wɪnʃi:ld] Windschutzscheibe
sanding block ['sændɪŋ ˌblɒk] Schleifklotz
sandpaper ['sænˌpeɪpə] Schmirgel- / Sandpapier
clear coat ['kliə ˌkəʊt] Klarlack
polishing compound ['pɒlɪʃɪŋ ˌkɒmpaʊnd] Polierpaste

C | MOT

MOT [ˌeməʊ'ti:] (britischer) TÜV
roadworthiness ['rəʊdˌwɜ:ðɪnəs] Verkehrstauglichkeit
exhaust emissions [ɪg'zɔ:st‿ɪˌmɪʃnz] Abgase
disc [dɪsk] *(hier:)* Plakette

appointment [ə'pɔɪntmənt] Termin
car registration number [ˌkɑ: redʒɪ'streɪʃn ˌnʌmbə] Autokennzeichen
courtesy car ['kɜ:təsi ˌkɑ:] Leihwagen

fuel cap ['fju:əl ˌkæp] Tankdeckel
exhaust pipe [ɪg'zɔ:st ˌpaɪp] Auspuffrohr
suspension [sə'spenʃn] Federung, Aufhängung
restraint system [rɪ'streɪnt ˌsɪstəm] Rückhaltesystem
latch [lætʃ] Riegel

Module 13 Trucks and lorries

truck [trʌk] LKW, Lastwagen
lorry ['lɒri] LKW, Lastwagen
LGV (large goods vehicle) [ˌeldʒi:'vi:, ˌlɑ:dʒ 'gʊdz ˌvɪəkl] Lastwagen
HGV (heavy goods vehicle) [ˌeɪtʃdʒi:'vi:, ˌhevi 'gʊdz ˌvɪəkl] Schwerlastwagen
PCV (passenger carrying vehicle) [ˌpæsɪndʒə 'kæriɪŋ ˌvɪəkl] Fahrzeug für die Personenbeförderung
commercial vehicle [kə'mɜ:ʃl ˌvɪəkl] Nutzfahrzeug
concrete mixer ['kɒŋkri:t ˌmɪksə] Betonmischer
tipper truck ['tɪpə ˌtrʌk] Kipplader
long-haulage lorry [ˌlɒŋ'hɔlɪdʒ ˌlɒri] Fernverkehrslastwagen
concrete ['kɒŋkri:t] Beton

A | Truck categories

weight category ['weɪt ˌkætəgri] Gewichtsklasse
special permit [ˌspeʃl 'pɜ:mɪt] Sondergenehmigung
articulated lorry [ɑ:'tɪkjəleɪtɪd ˌlɒri] Sattelschlepper
cab [kæb] Führerhaus; Taxi
trailer ['treɪlə] Anhänger
to be attached to sth. [bi‿ə'tætʃt tə ˌsʌmθɪŋ] an etw. befestigt sein
bend [bend] Kurve
GVM (gross vehicle mass) [ˌdʒi:vi:'em, ˌgrəʊs ˌvɪəkl 'mæs] Bruttogesamtgewicht
public ['pʌblɪk] öffentlich
pivot joint ['pɪvət ˌdʒɔɪnt] Drehgelenk

turning circle ['tɜ:nɪŋ ˌsɜ:kl] Wendekreis
driver drowsiness sensor [ˌdraɪvə 'draʊzɪnəs ˌsensə] Fahrerzustandserkennung
lane-departure warning system [ˌleɪndɪˌpɑ:tʃə 'wɔ:nɪŋ ˌsɪstəm] Fahrspurerkennungssystem
A-pillar ['eɪ ˌpɪlə] A-Säule
wheel arch ['wi:l‿ɑ:tʃ] Radlauf
rim [rɪm] Felge, Radkranz
tyre (BE) / tire (AE) [taɪə] Reifen

B | Interior cabin features

air vent ['eə ˌvent] Lüftungsschlitz
seat backrest ['si:t ˌbækrest] Sitzlehne
switch [swɪtʃ] Schalter
tray [treɪ] Ablage

passenger car ['pæsɪndʒə ˌkɑ:] PKW
to operate ['ɒpreɪt] fahren, bedienen
articulated lorry [ɑ:'tɪkjəleɪtɪd ˌlɒri] Sattelschlepper
hazard perception test [ˌhæzəd pə'sepʃn ˌtest] Gefahrenwahrnehmungstest

C | Truck dimensions

steering axle ['stɪərɪŋ‿æksl] Lenkachse
wheelbase ['wi:lbeɪs] Radstand
track [træk] Spurweite
tractor ['træktə] Zugmaschine
interaxle spacing [ˌɪntəræksl 'speɪsɪŋ] Zwischenachsabstand
drive axle ['draɪv‿æksl] Antriebsachse
kingpin ['kɪŋpɪn] Zugsattelzapfen
semi-trailer ['semɪˌtreɪlə] Sattelanhänger
measurement ['meʒəmənt] Maß, Abmessung
bumper ['bʌmpə] Stoßstange

Alphabetical word list

A

abbreviation Abkürzung 9
to accelerate beschleunigen 42
acceleration Beschleunigung 8, 58
accelerator (BE) / gas pedal (AE) Gas(pedal) 10, 66
access Zugang 69
to accommodate unterbringen, Platz bieten, berücksichtigen 11
according to entsprechend, gemäß, laut 25
accounts Buchhaltung 21
accuracy Genauigkeit 23
achievement Leistung 47
adhesive bonding Klebeverbindung 26
adjoining angrenzend 74
to adjust einstellen, anpassen 49
adjustable wrench Rollgabelschlüssel 16
advantage Vorteil 36
advert Anzeige 44
advice Rat 47
to affect Einfluss nehmen auf 71
to afford sich leisten 11
air conditioning Klimaanlage 11
air freight Luftfracht 69
airlock chamber Luftschleuse 34
air pollution Luftverschmutzung 67
air spring Luftfeder 42
air vent Lüftungsschlitz 90
aisle Fahrstraße, Gang 50
to align fluchten, ausrichten 49
all-season tyre Ganzjahresreifen 77
Allen key Innensechskantschlüssel, Inbusschlüssel 17
alloy Legierung 40
although obwohl 30
amazing erstaunlich, toll, beeindruckend 9
angle grinder Winkelschleifer, Flex 18
antifreeze Frostschutzmittel 81
anti-lock braking system Antiblockiersystem (ABS) 72
anti-skid traction control Traktionskontrolle und Antischleudersystem 73
A-pillar A-Säule 89
application Anstrich; Bewerbung 33, 44
to apply auftragen 33
appointment Termin 68
to appreciate etw. (wert)schätzen 37
apprentice Auszubildende(r), Lehrling 66
apprenticeship Ausbildung, Lehre 22
appropriate angemessen, passend 12
approximately ungefähr, etwa 25
area code Vorwahl 29
to argue diskutieren, streiten, argumentieren 8
argument Streit, Auseinandersetzung 71
articulated lorry Sattelschlepper 90
to assemble montieren, zusammenbauen 49
assembly Montage, Herstellung 45, 48
assembly shop Montagehalle 25
to assure zusichern, garantieren 26
attachment Anhang 38
audience Zuhörer *(Pl.)*, Publikum 61
automatic transmission Automatikgetriebe 11
autonomous cruise control system Tempomat 72
average durchschnittlich 11
to avoid vermeiden 71
award Auszeichnung, Preis 19
awkward peinlich, unangenehm 54

B

bank transfer Banküberweisung 38
base coat Basislack 32
beam (Licht-)Strahl 49
to be attached to sth. an etw. befestigt sein 89
beef Rindfleisch 55
to belong to gehören zu 23
bend Kurve 89
to be worth wert sein, lohnen 11
bill Rechnung 55
blade Schraubenblatt, Flügel 81
to blame sb. / sth. for sth. jmdm. / etw. die Schuld an etw. geben, jmdn. verantwortlich machen 71
blemish Makel 34
blind-spot detection Totwinkelerkennung 74
blister (Guss-)Blase 35
body Karosserie 18, 32, 49, 48
body in white Rohkarosserie, Karosseriegerippe 25
body language Körpersprache 61
body shell Rohbaukarosse 25, 32
body shop Karosseriewerkstatt, Rohbau 32
boiling point Siedepunkt 81
bolt (Schrauben-)Bolzen 17
bonnet (BE) / hood (AE) Motorhaube 10, 25, 37, 68, 87
boot (BE) / trunk (AE) Kofferraum 9, 24
boot lid Kofferraumdeckel 25
to bounce federn, hüpfen 42
B-pillar B-Säule 24, 73
bracket Klammer 82
brake Bremse 76
to brake bremsen 42
brake assist system Bremsassistenzsystem 72
brake booster Bremskraftverstärker 78
brake caliper Bremssattel 78
brake circuit Bremskreis 79
brake disc Bremsscheibe 78
brake drum Bremstrommel 79
brake fluid Bremsflüssigkeit 78, 85
brake horsepower (bhp) Pferdestärke (PS, am Rad gemessen) 58
brake pad Bremsklotz, Bremsbelag 85
brake pedal Bremspedal 78

cylinder bore Zylinderbohrung 59
cylinder head Zylinderkopf 21

D

damaged beschädigt 21, 69
to dampen dämpfen 42
dashboard Armaturenbrett 10
dead centre genaue Mitte 57
to deal with umgehen mit, sich befassen mit 68
to decide sich entscheiden 44
deformation zone Knautschzone 72
delay Verzögerung, Verspätung 28, 51, 59
delivery times Lieferzeiten 37
demand Nachfrage 72
dent Delle, Beule 86
department Abteilung 21
to depend on abhängig sein von 9
to deploy auslösen 73
to describe beschreiben 61
dessert Nachtisch, Nachspeise 55
to develop entwickeln 60
development Entwicklung 64
deviation Abweichung 25
device Gerät 21
dimension Maß, Abmessung 22
to dip eintauchen 32
disadvantage Nachteil 36
disc *(hier:)* Plakette 87
disc brake Scheibenbremse 78
to discontinue einstellen, beenden, auslaufen lassen 67
discount Rabatt, Preisnachlass 37
to dismiss nicht beachten 54
to dispatch versenden, verschicken 38
to disperse auflösen, zerstreuen 42
displacement Hubraum 58
distance Entfernung 64
to distribute vertreiben, verteilen 21
to dive (vorne) absinken; tauchen 42
to diversify aufteilen, verändern 19
DIY (Do-it-yourself) Heimwerken 38
dock Rampe 50
dolly Rollbrett 50
door handle Türgriff 10
door trim line Türverkleidungs-fließband 49
drawer Schublade 17
to drill bohren 17
drip Tropfen 35
drive axle Antriebsachse 40, 91
drive mechanism Antriebs-mechanismus 19
driver drowsiness sensor Fahrer-zustandserkennung 89
driveshaft Antriebswelle 43
drivetrain Antrieb(sstrang) 40, 76
droplet Tröpfchen 35
drum brake Trommelbremse 78
durable haltbar 63
dust cap Staubschutz 78
dust-free staubfrei 33

E

economical kostengünstig, wirtschaftlich 11
edge Kante, Rand 26
edge protection Kantenschutz 49
efficiently effizient 36
to eject herausschleudern 73
elevated erhöht 73
embarrassing peinlich 47
emission Emission, Ausstoß 64
to emit ausstoßen, emittieren 64
to employ beschäftigen 23
employee Beschäftigte(r), Angestellte(r) 21
employment contract Arbeitsvertrag 44
engine Motor 8
engineering Technik, Maschinenbau 19
enquiry Anfrage 68
to ensure gewährleisten, sicherstellen 25
environment Umwelt, Umgebung 8, 64, 80
environmentalist Umwelt-schützer(in) 8
essential sehr wichtig 68
estate car Kombi 73
to evaporate verdunsten 58
to exceed übersteigen, überschreiten 11
excuse Ausrede 71
executive car Vorstandswagen 9
exhaust Auspuff 65
exhaust emissions Abgase 87
exhaust manifold Abgaskrümmer 82
exhaust pipe Auspuffrohr 87
exhaust system Abgassystem 80
exhaust valve Auslassventil 57
existing bestehend 13
to expand sich ausdehnen 57
experienced erfahren 23
to explain erklären 68
to express zum Ausdruck bringen 20
to extend (sich) ausdehnen 42
extension Verlängerung 17
extension number Durchwahl 29
extinguisher Feuerlöscher 17

F

facility Einrichtung, Anlage 51
to fail scheitern, versagen 19
familiar vertraut 17
family name Nachname 15
family-owned in Familienbesitz 21
family-run in Familienbesitz, familiengeführt 37
fan Ventilator, Gebläse, Lüfter 81
fan belt Keilriemen 19
fatality Todesopfer 72
fault Fehler 45
faulty fehlerhaft, defekt, mangelhaft 69
to favour bevorzugen 9
feature Einrichtung, Ausstattung; Eigenschaft, Merkmal, Kennzeichen 9
fibre reinforced faserverstärkt 26
figures Nummern, Daten 31
file Feile 17
filler metal Schweißzusatz-werkstoff 26
filling station Tankstelle 64
first aid Erste Hilfe 17
to fix reparieren 28
flammable leicht entzündlich 67
floor pan Bodenwanne, Bodenblech 24
fluid Flüssigkeit 81
flush solution Spüllösung 83
flywheel Schwungrad 56
to fold falten 11

G

H

I

J

K

L

M

N

O

P

R

S

safety features Sicherheitsausstattung 72
sales pitch Verkaufspräsentation 60
saloon (car) Limousine 9, 25
salutation *(Brief:)* Anrede 37
sample Muster 37
to sand absanden, abschmirgeln 33
sanding block Schleifklotz 86
sandpaper Schmirgel-/Sandpapier 86
schedule Terminplan 84
scratch Kratzer 86
to screw schrauben 25
screwdriver Schraubendreher 16
to seal versiegeln, abdichten 33, 49
seam Saum 33
seat backrest Sitzlehne 90
seat belt Sicherheitsgurt 72
seat belt force limiter Gurtkraftbegrenzung 74
seat belt tensioner Gurtstraffer 73
seat-occupant sensing Sitzbelegungsfunktion 74
self-piercing rivet selbststanzende Niete 26
semi-trailer Sattelanhänger 91
sense Sinn, Bedeutung 11
service Dienstleistung 60
service mechanic Kfz-Service-mechaniker(in) 76
service record Service-Scheckheft 84
shaft Welle 56
to shape formen, prägen 17
to share teilen 9
to sharpen schleifen, schärfen, anspitzen 19
sheet metal Blech 25
shift Schicht 22
to shift schalten 58
shiny glänzend 62
shock absorber Stoßdämpfer 41
side curtain Seitenairbag 73
side order Beilage 55
side panel Seitenwand, Seitenteil, Seitenverkleidung 25, 49, 73
to sign unterschreiben 44
silencer Schalldämpfer, Auspufftopf 82
sill Schweller 24
size Größe 8
sketch Skizze, kurze Darstellung 17
skill Fähigkeit, Fertigkeit 19, 44
slack Lockerung, Durchhang 74
slide pin Bolzenführung 78
sliding seal Gleitdichtung 57
slot screwdriver Schlitzschraubendreher 17
smooth glatt, reibungslos 48
smoothness Glätte 25
snow tyre Winterreifen 77
socket Stecknuss 38
socket wrench Steckschlüssel 17
sodium azide Natriumazid (NaN_3) 74
softwood Weichholz 63
solution Lösung 28
to solve lösen 71
solvent Lösungsmittel 34
space Raum 9
spanner Schraubenschlüssel 16
spare tyre Ersatzreifen 17
spark Funken 74
spark plug Zündkerze 56
special permit Sondergenehmigung 89
specific bestimmt, spezifisch 69
speeding ticket Strafzettel für zu schnelles Fahren 66
speedometer Tachometer, Geschwindigkeitsmesser 10
to spell buchstabieren 47
to spin (sich) drehen 59
to split aufteilen 23
spotless fleckenfrei 33
spot welding Punktschweißen 26
spring Feder 42
to squat (hinten) absinken; (hin)hocken 42
to stall Motor abwürgen 43
to stamp stanzen, pressen 25
stamping shop Presswerk 24
starter Vorspeise 55
statement Aussage(satz) 25
state-of-the-art auf dem neuesten Stand (der Technik) 62
to steer lenken, steuern 41, 73
steering axle Lenkachse 91
steering column Lenksäule 49
steering wheel Lenkrad 74
stiff starr, fest, steif 26, 73
stock Bestand, Vorrat 38
to store verstauen 74
stranger Unbekannte(r) 15
strength Stärke 47
strut Federbein 42
stud Spike 78
sturdy robust 62
sub-assembly Baugruppe, Vormontage 25
subject line *(Brief:)* Betreffzeile 37
subsidiary Tochtergesellschaft 53
success Erfolg 68
suitable geeignet, passend 30
sulphur Schwefel 65
sum Betrag 39
summary Zusammenfassung 61
sump Ölwanne 56
supervisor Leiter(in), Kontrolleur(in) 47
supplier Lieferant 20
supply from stock ab Lager liefern 37
to support unterstützen 23
surplus überschüssig 69
to surround umgeben 24
suspension Federung, Aufhängung 11, 40, 76, 87
SUV (sport utility vehicle) Geländewagen 9, 41, 73
to swap tauschen 23
to swerve ausweichen 73
switch Schalter 90
synonym bedeutungsgleiches Wort, Synonym 17

T

tailgate Hecktür, Heckklappe 26
tail pipe Endrohr 82
tall hoch 73
tandem master cylinder Tandemhauptzylinder 79
tangle Durcheinander, Knäuel 17
tank lid Tankverschluss 25
target Ziel 9
task Aufgabe 22
tax Steuer 11
technician Techniker(in) 69
tension Spannung 54
term Begriff 72

U

V

W

Bildquellenverzeichnis

4 iStockphoto (Joachim Angeltun), Calgary, Alberta; **4** BMW AG, München; **4** Audi Medienzentrale @ picturesafe media/data/bank, Hannover; **5** iStockphoto (uchar), Calgary, Alberta; **6** LinguaTV GmbH, Berlin; **6** Video(s) supplied by BBC Motion Gallery, London; **6** ZDF, Mainz; **6** ZDF Enterprises GmbH, Mainz; **8** Daimler AG Medienarchiv, Stuttgart; **8** iStockphoto (Marin Tomas), Calgary, Alberta; **8** iStockphoto (Ermin Gutenberger), Calgary, Alberta; **8** iStockphoto (Joachim Angeltun), Calgary, Alberta; **10** iStockphoto (Trevor Smith), Calgary, Alberta; **10** iStockphoto (Anja Rabenstein), Calgary, Alberta; **11** Fotolia.com (Hauke-Chr. Dittrich), New York; **12** iStockphoto (daniel rodriguez), Calgary, Alberta; **12** Thinkstock (Stockbyte), München; **12** iStockphoto (David Jones), Calgary, Alberta; **12** iStockphoto (Helder Almeida), Calgary, Alberta; **14** iStockphoto (Dean Mitchell), Calgary, Alberta; **15** iStockphoto (4x6), Calgary, Alberta; **16** dreamstime.com (Leonid Yastremskiy), Brentwood, TN; **16** dreamstime.com (Christian Delbert), Brentwood, TN; **16** dreamstime.com (Nobilior), Brentwood, TN; **16** iStockphoto (Joe Belanger), Calgary, Alberta; **17** iStockphoto (bojan fatur), Calgary, Alberta; **18** dreamstime.com (Martin Kucera), Brentwood, TN; **18** iStockphoto (sumbul), Calgary, Alberta; **18** dreamstime.com (Radist), Brentwood, TN; **20** shutterstock (Dmitry Kalinovsky), New York, NY; **20** Thinkstock (BananaStock), München; **20** iStockphoto (sturti), Calgary, Alberta; **20** Thinkstock (iStockphoto), München; **22** shutterstock (Dmitry Kalinovsky), New York, NY; **23** iStockphoto (mediaphotos), Calgary, Alberta; **24** BMW AG, München; **25** BMW AG, München; **26** Daimler AG (Mercedes-Benz Cars Entwicklung), Sindelfingen; **28** iStockphoto (diego cervo), Calgary, Alberta; **28** iStockphoto (nolimitpictures), Calgary, Alberta; **28** Thinkstock (Wavebreak Media), München; **28** PantherMedia GmbH (Benis Arapovic), München; **30** iStockphoto (PeskyMonkey), Calgary, Alberta; **32** Dürr Aktiengesellschaft, Bietigheim-Bissingen; **32** Audi Medienzentrale @ picturesafe media/data/bank, Hannover; **33** AUDI AG, Ingolstadt; **34** Audi Medienzentrale @ picturesafe media/data/bank, Hannover; **35** Copyright: BASF; **36** iStockphoto (David Jones), Calgary, Alberta; **36** iStockphoto (Warwick Lister-Kaye), Calgary, Alberta; **36** shutterstock (Stephen Coburn), New York, NY; **36** shutterstock (rangizzz), New York, NY; **40** iStockphoto (Stefan Schulze), Calgary, Alberta; **41** iStockphoto (Izabela Habur), Calgary, Alberta; **42** iStockphoto (DangerJacobs), Calgary, Alberta; **43** Daimler AG Medienarchiv, Stuttgart; **44** shutterstock (Minerva Studio), New York, NY; **44** iStockphoto (Brian Jackson), Calgary, Alberta; **44** shutterstock (StockLite), New York, NY; **44** iStockphoto (Willie B. Thomas), Calgary, Alberta; **46** shutterstock (Goodluz), New York, NY; **48** shutterstock (Nataliya Hora), New York, NY; **50** iStockphoto (sturti), Calgary, Alberta; **50** dreamstime.com (Baloncici), Brentwood, TN; **50** iStockphoto (MotoEd), Calgary, Alberta; **51** shutterstock (Diego Cervo), New York, NY; **52** Thinkstock (Purestock), München; **52** iStockphoto (Chris Schmidt), Calgary, Alberta; **52** iStockphoto (fatihhoca), Calgary, Alberta; **52** iStockphoto (Clerkenwell_Images), Calgary, Alberta; **53** Thinkstock (BananaStock), München; **54** Thinkstock (Stockbyte), München; **55** iStockphoto (kristian sekulic), Calgary, Alberta; **56** Fotolia.com (motivation1965), New York; **56** dreamstime.com (Dalia Rukiene), Brentwood, TN; **56** dreamstime.com (Zoran Simin), Brentwood, TN; **56** shutterstock (apiguide), New York, NY; **57** shutterstock (Andrea Danti), New York, NY; **58** Fotolia.com (ferkelraggae), New York; **59** shutterstock (Iaroslav Neliubov), New York, NY; **60** iStockphoto (Clerkenwell_Images), Calgary, Alberta; **60** shutterstock (wavebreakmedia ltd), New York, NY; **60** Thinkstock (Purestock), München; **60** iStockphoto (Oleg Prikhodko), Calgary, Alberta; **61** iStockphoto (bluecinema), Calgary, Alberta; **63** iStockphoto (Joshua Hodge Photography), Calgary, Alberta; **64** Fotolia.com (Tom-Hanisch.de), New York; **64** Daimler AG Medienarchiv, Stuttgart; **64** dreamstime.com (Splosh), Brentwood, TN; **64** shutterstock (kylesmith), New York, NY; **65** Fotolia.com (joel_420), New York; **66** Daimler AG Medienarchiv, Stuttgart; **67** Daimler AG Medienarchiv, Stuttgart; **68** iStockphoto (YanLev), Calgary, Alberta; **68** iStockphoto (Steve Debenport), Calgary, Alberta; **68** iStockphoto (Sean Locke), Calgary, Alberta; **68** iStockphoto (Louis-Paul St-Onge), Calgary, Alberta; **69** iStockphoto (Sean Locke), Calgary, Alberta; **72** dreamstime.com (Flatscreen), Brentwood, TN; **72** shutterstock (Vereshchagin Dmitry), New York, NY; **72** Thinkstock (Digital Vision), München; **72** Honda Motor Europe, Offenbach am Main; **73** dreamstime.com (Jborzicchi), Brentwood, TN; **75** shutterstock (Ken Schulze), New York, NY; **76** iStockphoto (Alexandru Magurean), Calgary, Alberta; **76** iStockphoto (Brett Gage), Calgary, Alberta; **76** dreamstime.com (Pavel Mastepanov), Brentwood, TN; **76** iStockphoto (zubin li), Calgary, Alberta; **77** iStockphoto (bojan fatur), Calgary, Alberta; **78** QA International, Montreal; **79** Fotolia.com (Uwe Annas), New York; **80** shutterstock (Yu Lan), New York, NY; **80** iStockphoto (Donna Coleman), Calgary, Alberta; **80** shutterstock (KENCKOphotography), New York, NY; **80** dreamstime.com (Sixninepixels), Brentwood, TN; **82** iStockphoto (Michael Bodmann), Calgary, Alberta; **82** shutterstock (Christian Delbert), New York, NY; **82** shutterstock (Einar Muoni), New York, NY; **84** iStockphoto (uchar), Calgary, Alberta; **84** iStockphoto (sumbul), Calgary, Alberta; **84** iStockphoto (slobo), Calgary, Alberta; **84** iStockphoto (PeterTG), Calgary, Alberta; **85** iStockphoto (Brian Doty), Calgary, Alberta; **86** iStockphoto (egeeksen), Calgary, Alberta; **86** iStockphoto (Bruce Turner), Calgary, Alberta; **86** iStockphoto (manuel velasco), Calgary, Alberta; **88** iStockphoto (sumbul), Calgary, Alberta; **88** shutterstock (Bram van Broekhoven), New York, NY; **88** iStockphoto (Trevor Smith), Calgary, Alberta; **88** iStockphoto (Mike Clarke), Calgary, Alberta; **89** Daimler AG Medienarchiv, Stuttgart; **90** Thinkstock (Comstock), München; **91** shutterstock (Mechanik), New York, NY; **92** LinguaTV GmbH, Berlin; **93** LinguaTV GmbH, Berlin; **94** ZDF Enterprises GmbH, Mainz; **95** ZDF Enterprises GmbH, Mainz; **96** ZDF, Mainz; **96** ZDF Enterprises GmbH, Mainz; **97** Video(s) supplied by BBC Motion Gallery, London; **COVER** Getty Images (The Image Bank/Benedict Redgrove), München; **COVER** Getty Images RF (Cultura/Juice Images), München

Textquellenverzeichnis

101/102 Source: Powered Hand Tools - Pneumatic Tools - Basic Safety, http://www.ccohs.ca/oshanswers/safety_haz/power_tools/pneumat.html, Canadian Centre of Occupational Health and Safety (CCOHS), year of publication. Reproduced with the permission of CCOHS, 2012

Sollte es in einem Einzelfall nicht gelungen sein, den korrekten Rechteinhaber ausfindig zu machen, so werden berechtigte Ansprüche selbstverständlich im Rahmen der üblichen Regelungen abgegolten.

Wir bedanken uns bei Mitarbeitern der Firma Daimler AG für ihre Unterstützung bei der Erstellung dieses Buches.